inspirations

inspirations
the girls who dish

whitecap
Vancouver/Toronto

Edited by Elaine Jones
Proofread by Ann-Marie Metten
Cover design by Maxine Lea
Art direction by Roberta Batchelor
Interior design by Margaret Lee/Bamboo & Silk Design Inc.
Food photographs by Andrei Federov
Food styling by David Forestell
Author photograph by Greg Athans

Printed and bound in Canada

National Library of Canada Cataloguing in Publication Data
Main entry under title:
   The girls who dish!: inspirations

   Includes index.
   ISBN 1-55285-257-1

   1. Cookery. I. Barnaby, Karen.
TX715.6.G592 2001   641.5   C2001-910976-8

The publisher acknowledges the support of the Canada Council for the Arts and the Cultural Services Branch of the Government of British Columbia for our publishing program. We acknowledge the financial support of the Government of Canada through the Book Publishing Industry Development Program for our publishing activities.

Front cover photograph: Aunt Terry's Tomato Soup with Margaret's Chèvre Crostini (page 80)

CONTENTS

ENTRÉES (CONTINUED)

VEGETABLES

DESSERTS (CONTINUED)

appetizers

Roasted Mushroom Tapenade

**Makes 2 cups
(475 mL)**

They're not as exotic as Italian porcini, but roasting takes everyday mushrooms to another level (and rescues those a little past their prime hiding in the fridge). They emerge from the oven meaty and intense—making them perfect for mushroom soup or earthy sauces. Try this tapenade spread on a pizza crust topped with chèvre and grilled green onions or on my Crispy Potato Cakes (page 40). Diva at the Met serves its version on pesto-brushed baguette slices, topped with sun-dried tomato and fresh basil. Thank you, Michael Noble!

1 lb.	mixed mushrooms (button, cremini, portobello, shiitake, etc.)	455 g
3 Tbsp.	extra virgin olive oil	45 mL
2 tsp.	minced fresh rosemary	10 mL
1/4 tsp.	salt	1.2 mL
	freshly ground black pepper to taste	
1	ficelle or baguette, thinly sliced	1
1 cup	pitted black olives (California tinned variety)	240 mL
4	cloves garlic, minced	4
8	anchovy fillets	8
1 Tbsp.	capers, drained	15 mL
1/2–3/4 cup	extra virgin olive oil	120–180 mL

Preheat the oven to 425°F (220°C).

Remove pulpy or tough stems from the portobellos or shiitakes. Toss the mushrooms with the 3 Tbsp. (45 mL) olive oil, rosemary, salt and pepper. (Although mushrooms like salt for flavour, the olives and anchovies added later will contribute to the seasoning.) Spread on a cookie sheet and roast in the oven for 20 minutes. Some of the mushrooms should have toasted edges for the best flavour and some will be wrinkled, even dehydrated. Remove from the oven and cool.

GLENYS MORGAN

If serving the tapenade immediately, toast the bread slices in the hot oven. Arrange them in a single layer on a cookie sheet and toast them just enough to colour the edges.

Coarsely chop the mushrooms. Transfer to the food processor and add the olives, garlic, anchovy fillets and capers. Blend by pulsing the machine. With the motor running, add just enough olive oil in a steady stream to loosen the mixture, until it resembles the traditional olive tapenade—soft, juicy and easy to spread. Taste for salt and pepper and adjust the seasonings.

The rich flavour of any tapenade, generous in olive oil, is always best at room temperature. Make it ahead if desired, but let the tapenade come to room temperature before serving. Spoon the tapenade into a serving bowl with a teaspoon, just right for a nice small dollop as the crostini are passed.

tip: **Rosemary Toothpicks**

Rosemary toothpicks are great for garnishing appetizers and securing tea sandwiches. To make the toothpicks, cut sprigs of fresh rosemary about 2 to 3 inches (5 to 7.5 cm) long. Holding the sprigs by the tops, remove all but the top $1/2$ inch (1.2 cm) of needles (reserve the needles for cooking). The toothpicks can then be poked into your appetizers or sandwiches. Try serving Frikadeller (page 130) garnished with rosemary toothpicks. –MM

GLENYS MORGAN

Biscotti di Vino

Makes about 40 pieces

I love these crunchy, sweet and peppery biscuits. At the David Wood Food Shop in Toronto we used to sell them in their elegant tins from the DiCamillo Bakery in Buffalo, N.Y, but they are easily made at a fraction of the cost. They are great on their own, with the Tea-Scented Fresh Goat's Cheese (page 32) or with other cheese—especially a creamy Gorgonzola. I like to use a combination of black and white sesame seeds for a speckled effect.

2 cups	all-purpose flour	475 mL
1/4 cup	sugar	60 mL
1 1/2 tsp.	baking powder	7.5 mL
1 tsp.	sea salt	5 mL
1 tsp.	freshly ground black pepper	5 mL
1/2 cup	dry red wine	120 mL
1/2 cup	extra virgin olive oil	120 mL
1	egg white, beaten until foamy	1
4 Tbsp.	lightly toasted unhulled sesame seeds	60 mL

Preheat the oven to 350°F (175°C).

Stir together the flour, sugar, baking powder, salt and pepper in a large bowl. Add the wine and oil and stir with a wooden spoon just until smooth. The dough will be very stiff and a little oily.

Divide the dough into 4 pieces. Shape each piece into a 10-inch (25-cm) log. The logs will be lumpy looking. Flatten the logs slightly. Brush with egg white and sprinkle with 2 Tbsp. (30 mL) of the sesame seeds. Line 2 baking sheets with parchment paper and sprinkle the remaining sesame seeds evenly over the parchment.

Cut the logs into 1-inch (2.5-cm), slightly diagonal slices. Place the slices 1 inch (2.5 cm) apart on the baking sheets. Bake for 30-35 minutes or until lightly browned. Transfer to wire racks to cool. Store in a tightly covered container. These will keep well for 2-3 weeks.

KAREN BARNABY

Mussels with Peppers, Bamboo Shoots and Green Curry

My sister-in-law Taw taught me how to prepare green curry sauce. This recipe is a little tamer than hers—Taw likes her food spicy hot and would double the amount of curry paste and add several hot peppers. Thai food has become very popular and ingredients can now be found in most grocery stores. The Thai curry paste is available in yellow, green and red, with yellow being the mildest and red the hottest.

Serves 6

1 Tbsp.	vegetable oil	15 mL
2 tsp.	green curry paste	10 mL
1	14-oz. (398-mL) can coconut milk	1
2 Tbsp.	Thai fish sauce	30 mL
1 Tbsp.	granulated sugar or palm sugar	15 mL
1	kaffir lime leaf (optional)	1
1	small red bell pepper, cored and sliced into julienne	1
1	small green bell pepper, cored and sliced into julienne	1
1	8-oz. (227-mL) can sliced bamboo shoots, drained	1
72	fresh mussels, washed and debearded	72
1/2 cup	chopped fresh Thai basil or sweet basil	120 mL

Heat the oil in a large, wide, heavy pot over medium-high heat. Add the curry paste and cook for 1 minute, stirring constantly. Stir in the coconut milk, bring to a boil and cook for 2 minutes. Add the fish sauce, sugar, kaffir lime leaf, if desired, and red and green peppers. Simmer over medium-low heat for 7 minutes. Bring the sauce back to a boil over high heat and add the bamboo shoots and mussels. Cover the pot with a lid and cook until the mussels open, about 5-8 minutes. Stir in the basil. Serve immediately in warm bowls.

MARY MACKAY

Porcini Pâté

Serves 8 to 10

This is a quality pâté that does not cost an entire paycheque, but certainly tastes like it does. Porcini powder is available in specialty food stores, or you can purchase the dried mushrooms and grind them yourself in a food processor.

1 lb.	chicken livers	455 g
6 Tbsp.	unsalted butter	90 mL
1	medium yellow onion, diced	1
4 oz.	bacon	113 g
8	slices bacon	8
1 Tbsp.	Italian porcini powder	15 mL
1/4 cup	brandy	60 mL
	freshly ground black pepper	
10	dried bay leaves	10
3 Tbsp.	roasted unsalted pistachio nuts	45 mL
8	slices white bread	8

Preheat the oven to 250°F (120°C). Rinse the chicken livers under cool water and remove any visible membrane. Chop the livers, and set them aside.

Place the butter in a non-stick frying pan over medium heat. Add the onion and cook until it begins to soften. Add the 4 oz. (113 g) bacon and cook for about 5 minutes. Stir in the livers and continue to fry until they are cooked through, about 10–15 minutes.

Line a 7 1/2 x 3 3/4-inch (20 x 9-cm) loaf pan with the 8 slices of bacon, allowing about 1 inch (2.5 cm) of each slice to overlap the edge of the pan. Set aside.

Transfer the cooked liver mixture to a food processor and purée until smooth. Add the porcini powder, brandy and several twists of pepper. Pour the puréed mixture into the bacon-lined loaf pan. Place a layer of bay leaves on top of the pâté. Cover the top of the pâté with the overlapping bacon. Place the loaf pan in a hot water bath and bake for 1 1/2 hours.

Remove from the oven and cool on a rack. When the pâté is cool enough to handle, remove it from the pan, discarding the bacon and bay leaves. Place the pâté in a piping bag fitted with a star tip and pipe it into containers that suit your occasion. These could be Asian rice spoons, small butter dishes, mini soufflé pans or what-ever provides a mood and an occasion for your event. Garnish the top of each serving with a toasted pistachio nut.

Remove the crusts from the bread and cut each slice in half on the diagonal. Place on a baking sheet in a 250°F (120°C) oven and toast for 8-10 minutes. The toast points should be crisp and light brown, but still a little soft in the centre. Serve the pâté with the toast points.

quick bite: **Artichoke Crostini**

Drain and coarsely chop a small jar of artichoke hearts packed in oil or a can of water-packed artichoke hearts. Add a fair amount of finely grated Parmesan cheese, grated lemon rind, freshly ground black pepper to taste, and enough mayonnaise to bind the ingredients. Mix together, spread on thin slices of baguette and broil until bubbling. **–LS**

CAREN MCSHERRY-VALAGAO

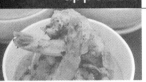

Shrimp and Goat Cheese Won Tons with Jade Sauce

Serves 4

This dish makes a wonderful start to a meal. The won tons are very light and the sauce is rich and flavourful. The won ton filling could also be prepared with fresh crab or lobster meat in place of the shrimp.

1	green onion, finely chopped	1
2 oz.	goat cheese, room temperature	57 g
6 oz.	raw shrimp	170 g
	pinch fine sea salt	
1/8 tsp.	freshly ground black pepper	.5 mL
20	round dumpling wrappers	20
1	beaten egg	1
1	clove garlic	1
1/4 tsp.	orange zest	1.2 mL
3/4 cup	packed fresh basil, chopped	180 mL
3/4 cup	packed fresh cilantro, chopped	180 mL
3/4 cup	packed fresh mint, chopped	180 mL
1 tsp.	rice wine vinegar	5 mL
1 tsp.	sesame oil	5 mL
1 tsp.	hoi sin sauce	5 mL
1/4 tsp.	Chinese chili sauce	1.2 mL
1/4 cup	chicken stock	60 mL
1/4 cup	whipping cream	60 mL
	fine sea salt and freshly ground black pepper to taste	
2 Tbsp.	vegetable oil	30 mL
2/3 cup	chicken or vegetable stock, divided	160 mL
	sprigs fresh cilantro	

In a medium bowl, stir together the green onion, goat cheese, shrimp, salt and pepper.

Line a baking sheet with parchment paper. Lay the dumpling wrappers out on a clean, dry surface. Brush the tops of the wrappers with egg. Place a heaping Tbsp. (15 mL) of shrimp filling in the middle of each wrapper. Fold the wrapper over top of the filling to form a semi-circle. Press down around the filling on the outside edges of each won ton to seal. Place the won tons on the baking sheet in a single layer and leave to dry.

To make the jade sauce, place the garlic, orange zest, basil, cilantro and mint in a food processor and blend until well combined, about 20 seconds. Scrape down the sides of the bowl. Add the rice wine vinegar, sesame oil, hoi sin sauce, chili sauce and $1/4$ cup (60 mL) of the chicken stock. Process until puréed, about 20 seconds.

In a small saucepan bring the herb purée to a boil over medium-high heat. Reduce the heat to medium and add the cream. Cook until slightly thickened, about 5 minutes. Season with salt and pepper. Keep the sauce warm while preparing the won tons.

Using two large frying pans over medium-high heat, heat 1 Tbsp. (15 mL) oil in each pan. When the oil is hot, place 10 won tons in each pan, being careful to not overcrowd them. Cook until the won tons begin to brown, about 2 minutes. Divide the $2/3$ cup (160 mL) stock evenly between the pans and cover the pans with lids. Reduce the heat to medium and steam for 1 minute.

Ladle the sauce into 4 heated pasta bowls. Transfer 5 won tons to each bowl of sauce. Garnish with fresh sprigs of cilantro and serve immediately.

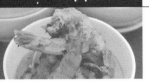

Lamb Pot Stickers with Cool Yogurt Sauce

Makes 36

Pot stickers have long been a favourite, usually using things that are leftover or easily available. This ground lamb filling opens a variety of options for the spicing. I have gone the Middle Eastern route, with sweet seasonings. (Pot sticker moulds are available at cooking supply stores everywhere for a few dollars. They give the pot stickers a crinkle-edge finish.)

1 Tbsp.	olive oil	15 mL
1 Tbsp.	unsalted butter	15 mL
2	large onions, sliced	2
2 Tbsp.	brown sugar	30 mL
1 lb.	lean ground lamb	455 g
1/4 tsp.	ground cloves	1.2 mL
1 tsp.	sea salt	5 mL
1 tsp.	freshly ground black pepper, preferably Tellicherry	5 mL
1 tsp.	ground cinnamon	5 mL
2 tsp.	ground cardamom	10 mL
1/2 tsp.	cayenne	2.5 mL
1/4 tsp.	freshly grated nutmeg	1.2 mL
2	peeled, boiled and mashed sweet potatoes	2
1	package thin won ton wrappers	1
1	large egg, beaten	1
1 recipe	Cool Yogurt Sauce	1 recipe

Heat a non-stick skillet over medium heat and add the oil and butter. Stir in the onions and cook until they are caramelized and a deep dark colour, about 20 minutes. Add the sugar and cook 10 minutes more, stirring frequently.

Meanwhile, heat another pan. Add the lamb and fry over medium-high heat until it begins to brown. Stir in the cloves, salt, pepper, cinnamon, cardamom, cayenne and nutmeg. Stir well to combine the flavours. Using a fork, break the lamb up so it is fine and even in texture, without any clumps of meat. Stir in the browned onions and sweet potato. Mix well to distribute the ingredients evenly. Taste and adjust the seasoning. Let cool.

To assemble, lay the won ton wrappers on your work surface. Place a generous tablespoon (15 mL) of the filling in the centre of each wrapper. Brush the edges of the wrapper with the egg. Fold the wrapper over and press the edges shut. Repeat until all the filling is used up.

Heat a non-stick 10-inch (25-cm) pan with $1/2$ cup (120 mL) water. Add the won tons to the pan and steam them until the water evaporates. Add a small amount of oil and fry the pot stickers for about 1 minute on each side, until they are golden brown. Serve with the yogurt sauce.

Cool Yogurt Sauce

Makes 2 cups (475 mL)

2 cups	Balkan-style yogurt	475 mL
$1/2$	lemon, juice only	$1/2$
2 tsp.	chopped fresh mint	10 mL
1 tsp.	ground cardamom	5 mL
1 tsp.	sea salt	5 mL

Mix all the ingredients together.

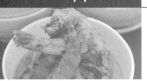

Vegetable Spring Rolls
with Spicy Dipping Sauce

Makes 36 rolls

Spring rolls are a bit time-consuming to make, but the effort is well worth it—especially when you receive accolades from friends and family.

6	dried Chinese mushrooms	6
1/2	package cellophane noodles	1/2
4 Tbsp.	peanut oil	60 mL
1 Tbsp.	sesame oil, preferably Kadoya	15 mL
4	green onions, thinly sliced	4
2	large shallots, sliced	2
3	cloves garlic, minced	3
1	large carrot, shredded	1
1/2 cup	chopped cilantro	120 mL
1	red bell pepper, julienned	1
1 tsp.	sugar	5 mL
2 tsp.	fish sauce	10 mL
2 Tbsp.	oyster sauce	30 mL
1	package 8-inch (20-cm) round rice papers	1
1 recipe	Spicy Dipping Sauce	1 recipe

Soak the mushrooms in hot water for 30 minutes. Squeeze them dry and cut off the woody stem. Julienne the mushrooms and set aside.

Soak the noodles in enough hot water to cover for 30 minutes. Drain, squeeze them dry in a kitchen towel and cut them into 1-inch (2.5-cm) lengths.

Heat 2 Tbsp. (30 mL) of the peanut oil and the sesame oil in a medium wok or sauté pan over medium heat. Add the green onions, shallots and garlic. Fry for a few minutes until softened, but not browned. Add the noodles, reserved mushrooms, carrots, cilantro, red pepper, sugar, fish sauce and oyster sauce. Stir-fry for about 1-2 minutes. Remove from the heat and cool.

Soak the rice papers in water for about 2 minutes, until they are pliable. Remove them from the water and place on tea towels to absorb any excess moisture. Cut the rice paper into quarters (a pizza wheel works great for this). Place a tablespoon (15 mL) of filling along the curved edge. Fold the corners in, and roll it up, creating a tight, neat roll.

Heat the remaining 2 Tbsp. (30 mL) of peanut oil in a frying pan. Add the rolls and fry until crispy and browned on all sides. Serve warm or at room temperature with the dipping sauce.

Spicy Dipping Sauce

Makes $^2/_3$ cup (160 mL)

1	clove garlic, minced	1
1	$^1/_2$-inch (1.2-cm) piece fresh ginger	1
2 Tbsp.	brown sugar	30 mL
2 Tbsp.	fish sauce	30 mL
2 Tbsp.	lime juice	30 mL
2 Tbsp.	rice vinegar	30 mL
2 Tbsp.	sambal oelek	30 mL
2 Tbsp.	chicken stock	30 mL

Combine all the ingredients in a small pot. Place over low heat and heat through to dissolve the sugar. Transfer to a bowl and serve with the spring rolls.

quick bite: **Prosciutto Asparagus Appetizers**

Spread a piece of prosciutto with a thin layer of Boursin cheese. Cut a spear of asparagus in half and lay the two pieces side by side at the narrow end of the prosciutto. Roll it up and bake at 350°F (175°C) for 5–8 minutes. **–LS**

CAREN MCSHERRY-VALAGAO

Breadmaking has always fascinated and inspired me.

BREADMAKING HAS ALWAYS fascinated and inspired me. In my eight years at Terra Breads, I have developed several recipes with my co-workers. As the bakery grows we are constantly improving our recipes and creating new ones, but unlike a restaurant, we do not create specials on a daily basis: our products take longer to develop. So I look for other sources of inspiration for my home cooking and entertaining.

Travel, such as trips to the Napa Valley, and reading great food magazines, such as *Fine Cooking, Cook's Illustrated* and *Martha Stewart Living*, are always sources of new ideas. Creative ideas also come from dining out at local restaurants, and we have a great pool of talented chefs in Vancouver.

But my biggest inspiration is my mother. I would not say that Mom loves cooking: I have often heard her complain about having to prepare another meal. It is more her whole approach to food that inspires me. Mom used cookbooks by local chefs, such as Susan Mendelson's books and *The Best of Bridge*. When I was in my teens, she took me to cooking classes with chefs like Anne Milne and Diane Clement. And I remember the day my Mom hired a caterer, Barbara Alexander, one of the first students from the Pierre Dubrulle Culinary School. Barbara brought along her school textbook and told me how much fun it was to be at cooking school. From that day on, my career was launched.

MARY MACKAY

MARY MACKAY

Mom generously offers gifts of food to almost everyone she meets. She picks up bread from Terra, choosing flavours carefully for each person on her list: fig and anise for Aunt Helen, Italian cheese for Uncle Jule and pumpkin seed for the neighbour next door. She bakes everyone's favourites—lemon cake for my sister Tede and me, bran muffins for my brother-in-law Gregg, and date squares for my other brother-in-law Steve. She makes a point of always having the cupboard stocked with goodies when the grand-children visit.

A year ago Mom was hospitalized after being struck by a van. Hooked up to life support and with countless broken bones, she managed to use her fingers to count out to me the amount of sugar called for in her bran muffin recipe. God forbid that anyone should go without bran muffins while she was laid up. Mom is now back to baking the bran muffins on her own, and at age 74 she continues to inspire me.

Tomato, Cumin and Black Pepper Tortillas with Chèvre

Makes 8 tortillas

Flour tortillas are very quick and easy to make. Traditional flour tortillas call for water to moisten the dough, but you can create some great flavours by substituting different vegetable juices for the water. Try using these tortillas in Deb's Quesadillas with Shrimp & Avocado Relish (*The Girls Who Dish*, page 8).

2 cups	unbleached all-purpose flour	475 mL
1/2 tsp.	fine sea salt	2.5 mL
1/2 tsp.	ground cumin	2.5 ml
1/4 tsp.	freshly ground black pepper	1.2 mL
3 1/2 Tbsp.	vegetable oil	52.5 mL
1/2–3/4 cup	tomato juice, room temperature	120–180 mL
	additional all-purpose flour for shaping	
1 1/2 cups	chèvre	360 mL

In a medium mixing bowl, stir together the flour, salt, cumin and pepper. Using a fork or pastry blender, stir in the vegetable oil. Mix in 1/2 cup (120 mL) tomato juice. Use your hands to gather the dough into a ball. If the dough is too dry to form a ball, work in more tomato juice, 1 Tbsp. (15 mL) at a time. Briefly knead the ball of dough on a lightly floured surface until soft and smooth, about 1–2 minutes.

Divide the dough into 8 equal pieces and form into balls. With lightly floured palms, flatten the balls into disks about 2–3 inches (5–7.5 cm) in diameter. Cover with plastic wrap and let sit for 20–30 minutes.

On a lightly floured surface, use a rolling pin to roll out a tortilla to 8 inches (20 cm) in diameter. Set the tortilla aside, cover with plastic wrap and continue rolling out the remaining tortillas.

Heat a 9-inch (23-cm) cast-iron frying pan over medium-high heat. When the pan is very hot, place a tortilla in the centre of the pan and cook for 45 seconds. Using a spatula, turn the tortilla over and cook another 45 seconds. The tortilla should be speckled golden brown on both sides. Lower the heat to medium if the tortillas are getting too dark. Remove from the pan, wrap in a dry kitchen towel and keep warm. Continue to cook the remaining tortillas.

Spread 3 Tbsp. (45 mL) chèvre on each tortilla, leaving a $1/2$-inch (1.2-cm) margin at the edge. Fold the tortilla in half. Cook in a heavy frying pan over medium heat until the tortilla is hot and the cheese has melted slightly. Cut into wedges to serve.

quick bite: **Mary Mac's Muffins**

I hate to admit it, but I love the Egg McMuffins at McDonald's. I find it frustrating that they stop preparing them after eleven in the morning, because they are great at any time of the day. I now make my own version and enjoy them at breakfast, lunch and dinner.

To make 2 muffins, heat a small baking sheet in a 450°F (230°C) oven until hot. Place 4 slices pancetta on the hot baking sheet, and toast in the oven until crisp. Scramble 3 large eggs with butter and season with salt and freshly ground black pepper. Toast 2 English muffins and spread with butter. Fill each muffin with 2 slices of pancetta, half of the scrambled eggs and a couple of slices of your favourite cheese. –MM

Caramelized Pear and Prosciutto Pizza

Serves 6 to 8

The saltiness of the prosciutto with the sweetness of the pear is the most delicious combination of flavours imaginable. If the dough makes this a page-turner, buy a portion of pizza dough at any good bakery. Lulu's balsamic vinegar is available at speciality food stores. If you can't get it, substitute your favourite balsamic.

For the dough:

1 tsp.	sugar	5 mL
1/2 cup	warm water	120 mL
1 Tbsp.	dry yeast	15 mL
3 cups	unbleached all-purpose flour	720 mL
2 tsp.	sea salt	10 mL
1 Tbsp.	chopped fresh rosemary	15 mL
3 Tbsp.	extra virgin olive oil	45 mL
3/4 cup	warm water	180 mL

Dissolve the sugar in the 1/2 cup (120 mL) warm water. Sprinkle the yeast on top and let it proof for about 5 minutes.

Place the flour, salt and rosemary in the bowl of a food processor. Add the oil, proofed yeast and remaining 3/4 cup (180 mL) water. Pulse the machine a few times to mix the dough, then let the machine run until the dough forms a ball on the side of the bowl, about 1 minute.

Remove the dough to a floured surface and knead it for about 2 minutes. The dough should be smooth and elastic. Place the dough in a bowl and rub the top with a little olive oil. Cover and let rise for about 1 hour, or until double in size.

For the topping:

3 Tbsp.	unsalted butter	45 mL
4	ripe Anjou or Bosc pears, peeled and	
	sliced 1/4 inch (.6 cm) thick	4
3 Tbsp.	Lulu's fig-infused balsamic vinegar	45 mL
3 Tbsp.	freshly grated Parmesan cheese	45 mL
10-12	slices prosciutto di Parma, chopped	10-12
8 oz.	Gorgonzola cheese, crumbled	227 g
1	bunch fresh basil, chopped	1
	freshly ground black pepper to taste	

Preheat the oven to 425°F (220°C). Oil an 11 x 17-inch (28 x 43-cm) cookie sheet.

Melt the butter in a shallow frying pan; add the pears and vinegar. Turn the heat to high and cook the pears on both sides, for a total of about 3 minutes. Remove from the heat.

Turn the dough onto the prepared cookie sheet and press it evenly to the edges. Brush the surface with olive oil and sprinkle with the Parmesan cheese. Place the pears evenly over the dough. Sprinkle with the prosciutto and top with the crumbled Gorgonzola. Bake for about 15-20 minutes, or until the edges of the crust are golden brown. Remove the pizza from the oven. Sprinkle the chopped basil over the top and finish with a good grinding of pepper.

tip: **Sea Salt**

Salt is one of the most crucial ingredients in cooking, and the right kind of salt is a small investment in the difference between good and great! Sea salt, kosher salt and fleur de sel bring out the flavours of your food without adding the overtone of chemicals found in iodized table salt.

–LS

CAREN MCSHERRY-VALAGAO

Almond-Crusted Camembert
with Pear Cranberry Chutney

Serves 4 to 6

This is a great little appetizer. Serve it on a brightly coloured plate with rounds of French bread and an assortment of quality crackers. You can make the chutney a day or two ahead and keep it in the fridge; just be sure to bring it back up to room temperature before serving. I like to use dried cranberries, as they are intensely flavoured and don't release moisture during cooking.

3 Tbsp.	flour	45 mL
1	large egg	1
1/4 cup	toasted almonds, finely ground	60 mL
1/4 cup	dried bread crumbs, finely ground	60 mL
1/4 tsp.	freshly ground black pepper	2.5 mL
2	3-inch (7.5-cm) rounds Camembert cheese	2
3 Tbsp.	olive oil	45 mL
1 recipe	Pear Cranberry Chutney	1 recipe

Place the flour in a shallow dish. Break the egg into another shallow dish and beat lightly with a fork. Combine the ground almonds, bread crumbs and pepper in a third shallow dish.

Dredge each Camembert round in the flour, dip it in the egg mixture and then roll it in the crumb mixture, pressing gently until it is completely coated.

Heat the olive oil in a sauté pan over medium-high heat. When the oil is very hot but not smoking, fry the cheese rounds for 2 minutes. Reduce the heat to medium, gently turn the cheese and fry for 2 minutes more. Remove the cheese to paper towels to drain.

To serve, overlap the Camembert in the centre of a dinner plate. Top with the chutney and grind black pepper around the edge of the plate.

Pear Cranberry Chutney

Makes 1 cup (240 mL)

3 Tbsp.	water	45 mL
3 Tbsp.	sugar	45 mL
2 Tbsp.	finely diced red onion	30 mL
3 Tbsp.	red wine vinegar	45 mL
2	pears, peeled, cored and thinly sliced	2
1/2 cup	dried cranberries	120 mL
1 Tbsp.	honey	15 mL

In a medium sauté pan over medium-high heat combine the water and sugar. Bring to a boil, lower the heat to medium and cook until the sugar is lightly caramelized; it should be a light golden colour. Add the onion and continue to cook for 3–4 minutes. Add the vinegar and pears, turn the heat to medium-low and simmer gently for 3–4 minutes. Add the cranberries and honey and cook gently for 3–4 minutes. Remove from the heat and set aside to cool.

tip: **No-Fuss Oil**

When a recipe calls for brushing with oil, an easy way to avoid brushes and bowls is to use the handy oil sprayers, a no-fuss way to apply oil. When making crisp pita triangles, I simply spray the pita breads, cut them into triangles and toast them in the oven. **–CMV**

Tea-Scented Fresh Goat Cheese

Serves 4

This recipe is a combination of inspirations—David Wood's fabulous Salt Spring Island goat cheese and T Tearoom's wonderful teas. Cheese marinated in jasmine, Earl Grey or a fruit tea would be good with fruit, while the nutty flavour of Genmai Cha or the smoky flavour of Lapsang Souchong would be good in a salad.

1	soft unripened goat's cheese, approximately 4 oz. (115 g)	1
2 Tbsp.	fragrant tea (natural fruit tea, Lapsang Souchong, Genmai Cha, jasmine, Earl Grey, etc.)	30 mL

Wrap the cheese in a single layer of cheesecloth with very little overlap.

Place a piece of plastic wrap that is large enough to completely wrap the cheese on your work surface. Spread 1 Tbsp. (15 mL) of the tea in a band lengthwise down the middle of the wrap. Place the curved edge of the cheese in the tea at one end of the wrap and tightly roll the cheese in the wrap, making sure that the tea is covering the side of the cheese all the way around. Place the cheese with one of the flat sides up and sprinkle half the remaining tea over the cheese. Enclose it tightly with the wrap. Repeat on the remaining side. You should end up with a piece of cheese that is almost completely covered in tea, tightly wrapped in plastic. Refrigerate for 24 hours.

Remove the plastic wrap and cheesecloth from the cheese. Wrap in plastic wrap to store the cheese if you are not serving it right away. Like all cheese, this one is better at room temperature.

KAREN BARNABY

Curried Cauliflower
Soup with Apple
and Almond Relish
(page 88)

Pan-Seared Swordfish
with Mediterranean
Relish (page 112)

Bocconcini
with Kalonji Seeds,
Cilantro and Tomato
(page 71)

Fiery Thai Beef with
Herb Salad and
Cooling Cucumber
Pickle (page 134)

Camembert Fondue

No need to pull out that old fondue pot, this fondue comes with its own edible bowl—a hollowed-out round loaf of sourdough bread. I use a local Camembert goat cheese from Salt Spring Island and serve the fondue with sliced baguette, fruit and crackers.

Serves 8 to 10

1	1-lb. (455-g) round loaf of bread, preferably sourdough	1
1	wheel Camembert cheese, about 4 inches (10 cm) in diameter	1
	sprigs fresh thyme	

Preheat the oven to 450°F (230°C).

Using a serrated knife, slice off the top of the loaf, a quarter of the way down. Slice the top into small, bite-size pieces and set aside. Using the wheel of Camembert as a guide, remove enough of the interior breadcrumb to fit the cheese inside the loaf of bread. Use your fingers to gently pull out the interior of the loaf. Work your way around the inside edges, being careful not to make holes in the crust. Turn the loaf over and gently score the bottom crust with the serrated knife, making sure not to cut all the way through. Place the wheel of Camembert inside the hole in the loaf of bread.

Wrap tin foil around the outside of the loaf of bread, leaving the cheese exposed. Place the loaf on a baking sheet and bake in the oven for 12 minutes. Remove the tin foil and bake another 3 minutes, until the cheese is fully melted. Baking times may vary with the age and temperature of the cheese. Keep an eye on the loaf while it is baking so it does not burn.

Place the cheese fondue on a platter and serve immediately. Garnish with fresh thyme sprigs and extra pieces of sourdough bread for dipping.

MARY MACKAY

Seared Scallops with Apple Potato Pancakes and Chanterelles

Serves 6

This dish combines some of my favourite flavours. You can toss a small amount of salad greens with the chanterelles and scallops to turn it into a light entrée. The potato cakes should be assembled just prior to cooking as the potatoes will discolour if they are left to sit. Clean chanterelle mushrooms with a cloth or a small brush; do not wash them as they will absorb moisture. If the mushrooms are large, tear them into bite-size pieces.

4 Tbsp.	olive oil	60 mL
1/2 cup	minced white onion	120 mL
1	Granny Smith apple	1
4 Tbsp.	white wine	60 mL
1 tsp.	salt	5 mL
1/2 tsp.	freshly ground black pepper	2.5 mL
1 1/2 lbs.	Yukon gold potatoes	680 g
2 Tbsp.	all-purpose flour	30 mL
2 Tbsp.	butter	30 mL
12	large scallops	12
1 lb.	chanterelle mushrooms	455 g
2 Tbsp.	apple cider	30 mL
1 tsp.	honey	5 mL

Preheat the oven to 325°F (165°C).

Heat 2 Tbsp. (30 mL) of the olive oil in a sauté pan over medium-high heat and add the onions. Reduce the heat to medium and cook for 5 minutes, stirring occasionally, until the onion is translucent and lightly coloured.

Peel and core the apples and cut them into a fine dice. Add apples to the pan with 2 Tbsp. (30 mL) of the white wine. Add the salt and pepper and continue to cook over medium heat until all the liquid has evaporated. Remove the mixture from the pan and set it aside to cool.

Grate the potatoes, squeeze out the excess moisture and place them in a medium bowl. Add the onion-apple mixture to the grated potatoes and stir. Add the flour and mix well. Shape the potato mixture by hand into 6 patties, approximately 3 inches (7.5 cm) across and $1/2$ inch (1.2 cm) thick.

In a non-stick pan, heat 1 Tbsp. (15 mL) of the olive oil with 1 Tbsp. (15 mL) of the butter over medium-high heat. When the oil is hot, reduce the heat to medium and place the pancakes in the pan. Do not overcrowd the pan. Cook the pancakes until crisp, about 5 minutes. Turn the pancakes and cook for 5 minutes more. Place them in the oven to keep warm.

Using the same pan, heat the remaining 1 Tbsp. (15 mL) of olive oil over medium-high heat. When the oil is hot add the scallops and cook 1-2 minutes per side, until they are just cooked through. Place them in the oven to keep warm. Add the chanterelle mushrooms to the pan and season to taste with salt and pepper. Sauté, stirring occasionally, for 5-6 minutes, until they are lightly browned. Remove the mushrooms with a slotted spoon and place in the warm oven.

Deglaze the pan with the apple cider. Turn the heat to low, and use a wooden spoon to scrape the pan. Add the honey and remaining 2 Tbsp. (30 mL) of white wine. When it's heated through, remove the pan from the heat and stir in the remaining 1 Tbsp. (15 mL) of butter.

To serve, set each pancake in the centre of a small plate. Top with chanterelle mushrooms and scallops and drizzle with the pan juices.

Steamed Clams with Tomatoes, White Wine and Fresh Thyme

Serves 4

Clams are available year round and this appetizer can be put together in minutes. Serve the clams in large shallow bowls with plenty of fresh bread for dipping. You can make the dish more substantial by adding diced, blanched potatoes.

2 Tbsp.	olive oil	30 mL
3	shallots, diced	3
1	clove garlic, minced	1
3 lbs.	fresh clams, scrubbed	1.35 kg
1 tsp.	freshly squeezed lemon juice	5 mL
1/2 cup	white wine	120 mL
1/4 cup	whipping cream	60 mL
	salt and freshly ground black pepper to taste	
1 tsp.	fresh thyme leaves	5 mL
2	medium tomatoes, seeded and diced	2
2 Tbsp.	softened butter	30 mL
2 Tbsp.	chopped chives	30 mL

Heat the olive oil in a large sauté pan over medium heat. When the oil is hot add the shallots and garlic; sauté for 2 minutes. Add the clams, lemon juice, wine and cream. Season lightly with salt and pepper.

Increase the heat to medium-high. Cover the pan with a lid and cook for 6–8 minutes, or until all the shells have opened. Remove the lid and add the thyme and tomatoes; cook for 2 more minutes. Remove the pan from the heat and stir in the softened butter and chives.

Chai Spiced Nuts

As you may have noticed, I have a real love affair going on with chai! This is something I came up with one Christmas. The coconut adds a Far Eastern flair. It is an easy and much-appreciated gift. I use T Tearoom's Herbal Spice Chai, but any chai without tea leaves or chai masala that can be purchased in East Indian grocery stores can be used.

Makes 3 1/2 cups (840 mL)

3 Tbsp.	sugar	45 mL
1/2 tsp.	cayenne	2.5 mL
1/2 tsp.	salt	2.5 mL
1 1/2 tsp.	Herbal Spice Chai	7.5 mL
1	egg white	1
1 cup	whole raw almonds	240 mL
1 cup	whole raw cashews	240 mL
1 cup	whole raw pecan halves	240 mL
1/2 cup	unsweetened, shredded coconut	120 mL

Preheat the oven to 350°F (175°C).

Combine the sugar, cayenne, salt and chai. In a separate bowl, beat the egg white with a whisk until foamy but not stiff. Add the nuts and coconut. Stir to coat with the egg white. Add the chai mixture and stir until evenly blended. Spread out in a single layer on a parchment-lined baking sheet and bake for 10 minutes. Stir the nuts with a spoon and bake for 10 minutes longer. Cool completely in the pan. Keeps for 2 weeks at room temperature, tightly covered.

KAREN BARNABY

Coconut-Crusted Prawns
with Mango Tango Sauce

Serves 4

At Lesley Stowe Fine Foods we are constantly challenged to come up with new hors d'oeuvre ideas—and these prawns a standby that keeps resurfacing. It is definitely worth hunting out the unsweetened thread coconut for an interesting crust on the outside.

1	egg yolk	1
1 cup	cold water	240 mL
1 cup	flour	240 mL
1 cup	shredded unsweetened coconut	240 mL
1 Tbsp.	cornstarch	15 mL
2	egg whites	2
1 tsp.	salt	5 mL
2 Tbsp.	minced chives (or cilantro)	30 mL
4 cups	safflower or grapeseed oil	950 mL
1 1/2 lbs.	prawns, peeled, tails left on	680 g
1/2 cup	cornstarch	120 mL
1 recipe	Mango Tango Sauce	1 recipe

Stir together the egg yolk and water. In a separate bowl, stir together the flour, coconut and the 1 Tbsp. (15 mL) cornstarch.

Whip the egg whites to stiff peaks. Combine the yolk mixture with the flour mixture. Fold in the egg whites, salt and chives.

In a deep, heavy straight-sided pot or Dutch oven, heat 3 inches (7.5 cm) of safflower or grapeseed oil to 350°F (175°C). Coat the prawns with the 1/2 cup (120 mL) cornstarch and shake off the excess. Holding each prawn by the tail, dip it into the batter. Do not coat the tail. Let the excess drip off. Place the prawns in the hot oil and deep fry for 2–3 minutes. Drain the prawns on paper towels and serve warm with the dipping sauce.

Mango Tango Sauce

Makes 1 cup (240 mL)

1 cup	sour cream	240 mL
2 Tbsp.	mango chutney, minced to a fine consistency	30 mL
1 Tbsp.	honey	15 mL
1 tsp.	cumin	5 mL
2 tsp.	grated fresh ginger	10 mL

Combine all the ingredients. Refrigerate until needed.

quick bite: **Tea and Almonds**

My current favourite late-night snack is butter-toasted almonds with balsamic vinegar and kosher salt served alongside a fragrant cup of Herbal Spice Chai. Here's how to do it . . .

Spread raw, unblanched almonds in a single layer on a microwavable plate. Toss on a lump of unsalted butter and microwave on high for a minute. Stir the almonds and microwave for another 30 seconds. Stir again and cook for 30 seconds more. Check an almond to see if it is roasted enough. If not, continue in 30-second increments until they're roasted. Remove the nuts and sprinkle them with kosher or sea salt and a drizzle of good balsamic vinegar. Let cool a bit before munching.

For the chai, I just run a teaspoon or so of chai through my espresso machine as if making an Americano. Add sweetener and cream to taste.
–KB

Crispy Potato Cakes

**Makes
6 to 8 cakes**

Everyone's mad for mashed, but this version takes the cake! Stylishly professional looking (made with the help of a few round cookie cutters), one spoonful says comfort food. The tower of potato with its crispy top and bottom, creamy white sides and hint of horseradish is the perfect base for a show-stopping appetizer. Some of my favourite ways to top the cake: sour cream and caviar with chives; slices of smoked salmon or trout with crème fraiche and herbs; smoked trout mousse with dill sprigs; a dollop of Roasted Mushroom Tapenade (page 12) with mascarpone and cracked pepper.

1 lb.	baking potatoes, such as russet or Idaho (about 2 large)	455 g
2 Tbsp.	finely grated onion	30 mL
1 Tbsp.	horseradish (or more to taste)	15 mL
2 Tbsp.	flour	30 mL
1/2 tsp.	salt	2.5 mL
	freshly ground white pepper to taste	
	generous pinch nutmeg	
2	extra large eggs, separated	2
	vegetable spray	
3-5 Tbsp.	vegetable oil or clarified butter (page 139)	45-75 mL

Scrub the potatoes and place in a saucepan. (Cook baking potatoes whole, in their skins, keeping the inside dry and fluffy.) Cover with water and bring to a boil; to avoid splitting the skin, reduce the heat to a high simmer. Cook until fork tender. Cool the potatoes, peel and mash. For the best texture, use a potato ricer or food mill. Combine with the grated onion, horseradish and flour. Season with salt, pepper and nutmeg. Taste and adjust the seasonings before folding in the egg yolks.

In a large bowl, beat the egg whites until firm; they should hold a peak on the whisk. Use a large spatula to fold the beaten egg whites into the potato mixture.

Prepare round 2- to 3-inch (5- to 7.5-cm) cookie cutters or specialty rings by spraying the inside of each ring with vegetable spray. Heat a non-stick pan over medium-high heat and add enough oil to cover the bottom of the pan. Place several rings in the pan without crowding them. Spoon in the filling until it's level with the top of the ring, and gently even it out with the back of the spoon.

Cook the first side of the cake longer, adjusting the heat so the cake develops a nice brown crust. After the cakes have been cooking for several minutes (undisturbed!) use a spatula to lift them and check the colour. Once the bottom is set and nicely browned, use two spatulas to flip the cakes over.

Reduce the heat to medium-low to cook the second side. A good sign is when the potato "soufflés" out of the ring. Work in batches, adding oil as necessary. Hold or rewarm the cakes on a baking sheet in a 150°F (65°C) oven. At this point you can slide the rings off to make the next batch.

If you'd like a natural presentation—like fluffy clouds—or if you don't have specialty rings, the cakes can be made free-form. It's best to make them thick, so they stay moist inside. Spoon 3 Tbsp. (45 mL) of the batter into the pan, mounding the soft mixture about 2 inches (5 cm) high without flattening the top. Let the cakes brown nicely on the edges before turning them over. Cook about 2–3 minutes on each side.

For the fluffiest potato cakes—shaped or free-form—serve them right from the pan or keep the resting time in the oven short. If they do sit, they will be more dense, but just as tasty.

GLENYS MORGAN

Salmon and Potato Napoleons
with Seasonal Greens in Citrus Vinaigrette

Serves 8

I am often influenced and inspired by the food that I eat while on vacation. A small, off-the-beaten-track restaurant in Portland caught my attention with this stacked creation.

2 cups	peanut or grapeseed oil	475 mL
24	won ton wrappers, cut in half on the diagonal	24
3	russet potatoes	3
4 Tbsp.	unsalted butter	60 mL
1/4 cup	whole milk	60 mL
1/2 tsp.	salt	2.5 mL
	freshly ground black pepper to taste	
1 Tbsp.	wasabi mustard	15 mL
4 cups	seasonal salad greens	950 mL
16	slices cold smoked salmon, about 3/4 oz. (20 g) each	16
1/2 cup	crème fraiche (page 185)	120 mL
	fresh whole cilantro leaves	
3/4 cup	Citrus Vinaigrette	180 mL

In a heavy, straight-sided skillet, heat the oil to 350°F (175°C). Add the wrappers 2 at a time and cook for 30 seconds, or until golden. Drain and cool.

Peel the potatoes and cut them in quarters. Place in a pot, cover with cold water and bring to a boil. Simmer for 20–30 minutes. Drain the potatoes and pass them through a ricer. Heat the butter and milk together and add them to the potatoes. Add the salt, pepper and wasabi.

To assemble individual salads, place a won ton wrapper on a plate. Place a spoonful of potato in the centre of the wrapper. Place a few salad greens on top, drape with a slice of smoked salmon and top with a won ton wrapper. Repeat the layers, ending with a won ton wrapper. Garnish with crème fraîche and cilantro. Dress the remaining salad greens with the citrus vinaigrette and serve.

Citrus Vinaigrette

Makes 1 cup (240 mL)

2 Tbsp.	lemon juice	30 mL
2 Tbsp.	orange juice	30 mL
1 tsp.	freshly grated lime rind	5 mL
1 tsp.	Dijon mustard	5 mL
3/4 cup	extra virgin olive oil	180 mL
	salt and freshly ground black pepper to taste	

Whisk together the lemon juice, orange juice, lime rind and mustard. Keep whisking and gradually add the olive oil. Season with salt and pepper. This vinaigrette will keep for 1 week in the fridge.

Cranberry Onion Confit
and Stilton Crostini

Makes 1 1/2 cups
(360 mL) confit

My friend Ann Kirsebom is the chef behind the bright, bold Tequi-Lime BBQ Sauce, which has devotees from the Texas barbecue belt to hibachis beachside in Kits. It's a bottled version of Ann's style as a caterer, with Asian influences, Southwestern heat and classics given an artistic twist. Her confit, with its dark berry flavours and the tang of cheese, is my idea of being catered to!

3 Tbsp.	olive oil	45 mL
3 cups	very thinly sliced red onion	720 mL
1/2 cup	sun-dried cranberries	120 mL
4 Tbsp.	demerara sugar	60 mL
4 Tbsp.	balsamic vinegar	60 mL
1	baguette, thinly sliced and toasted	1
4 oz.	Stilton (or Roquefort)	113 g

In a saucepan with a cover, heat the olive oil over medium heat. Add the onions and toss to coat. Cover the saucepan and very gently cook the onions until they melt, almost losing their shape.

Stir in the cranberries, sugar and vinegar. Stir to mix and turn the heat down very low. Let the juices evaporate, creating a marmalade texture. Since onions vary in sweetness, taste for a nice balance of sweet and sour. Adjust to your taste buds with a little more sugar or vinegar.

Use a teaspoon to spoon the confit on the toasted baguette slices. Use a sharp paring knife (or a small spoon for softer cheese, such as Boursin) to add a small shaving or dab of cheese.

The cheese may also be spread on the baguette slices with the confit on top. Serve at room temperature for the best flavour.

Minted Eggplant, Feta and Shallot Platter

This dish was one part "market inspiration" and one part tired chef. It was a Saturday afternoon in August. I had already put in a full day's work overseeing a dozen or so parties with guests totalling seven hundred. I had been invited to a potluck party, which meant there was still more cooking to do. A stroll through my neighbourhood market uncovered the biggest, most "aubergine"-coloured eggplants I had ever seen. I headed home with two of them, too tired to shop for anything else. I fired up my barbecue and while the eggplants grilled, I foraged through my larder and snipped mint from my balcony herb garden. In the end, the dish paid homage to southern France and Italy. It was a hit.

Serves 10 to 12

2	eggplants	2
3 Tbsp.	extra virgin olive oil	45 mL
3/4 tsp.	sea salt	4 mL
1/2 tsp.	freshly ground black pepper	2.5 mL
2/3 cup	lightly packed mint, chopped	160 mL
1 cup	crumbled feta	240 mL
1/4 cup	finely chopped shallots	60 mL
1 1/2 Tbsp.	red wine vinegar	22.5 mL
4 Tbsp.	extra virgin olive oil	60 mL
6	pieces pita bread, quartered	6

Preheat the barbecue, grill pan or broiler.

Slice the eggplants crosswise into 1/2-inch (1.2-cm) pieces. Brush with the 3 Tbsp. (45 mL) olive oil. Season with salt and pepper. Grill or broil until well browned and soft throughout (this is the key to a good result). Set the eggplant aside to cool.

Arrange half the eggplant slices on a platter..Sprinkle with half the mint, feta, shallots, red wine vinegar and remaining olive oil. Arrange the rest of the eggplant and sprinkle with the remaining mint, feta, shallots, vinegar and olive oil. Go to the party. Serve at room temperature with pita bread.

MARGARET CHISHOLM

salads

Portobello Mushroom Salad
with Gorgonzola Dressing

Serves 4

I have never been a big fan of mushrooms, but for some reason I tried them on a salad with blue cheese one day and voilà! I saw the light. Well, maybe not the light, but the combination is extremely delicious. This salad makes a great appetizer, or you can serve it with a steak as a main course.

4 oz.	Gorgonzola cheese, rind removed	113 g
1 cup	sour cream	240 mL
1/4 cup	mayonnaise	60 mL
1 Tbsp.	red wine vinegar	15 mL
1 tsp.	minced garlic	5 mL
	salt and freshly ground black pepper to taste	
1	red bell pepper	1
4	portobello mushrooms, 3–4 oz. (85-113 g) each	4
	extra virgin olive oil	
2	romaine hearts	2

Crumble the cheese and mash it to a paste. Slowly stir in the sour cream, then the mayonnaise, vinegar and garlic. Season with salt and pepper.

Place the bell pepper on a hot grill and grill until blackened on all sides. Cool, then peel off the blackened skin. Remove the stem and seeds and cut the pepper into 4 pieces.

Remove the stems from the mushrooms and scrape out the gills with a small spoon. Score the tops in a cross-hatch pattern. Coat both sides of the mushrooms with oil, and season with salt and pepper. Preheat the broiler to high. Place the mushrooms on a baking sheet, stem side up. Broil for 2-3 minutes on each side until cooked through.

Cut the romaine hearts in half and divide among 4 plates. Drizzle with the dressing. Top with a mushroom and a piece of red pepper.

KAREN BARNABY

Greens with Crispy Prosciutto and Cambozola Croutons

A friend/client first told me about using crispy prosciutto instead of bacon in her salads. It adds a great salty component and is fabulous paired with Cambozola croutons and sweet, port-soaked cherries.

Serves 6

8	slices prosciutto	8
8	slices baguette	8
8 oz.	Cambozola	227 g
1	head red leaf lettuce	1
1	bunch watercress	1
1/2	head romaine lettuce	1/2
1/2 cup	dried cherries	120 mL
1 cup	Late-bottled Vintage Port	240 mL
2 tsp.	grainy Dijon mustard	10 mL
1/4 cup	good-quality red wine vinegar	60 mL
1 cup	extra virgin olive oil	240 mL
	sea salt and freshly ground black pepper to taste	

Preheat the oven to 375°F (190°C). In a large sauté pan over medium-high heat, cook the prosciutto in a single layer until crispy. Drain and set aside.

Lightly toast the baguette slices. Place them on a baking sheet and top each with 1 oz. (30 g) of Cambozola. Set aside.

Wash, dry and break up the leaves of the red leaf lettuce, watercress and romaine. Place in a large bowl. In a small saucepan over medium-high heat, simmer the cherries and port until the cherries are plump and approximately 1 Tbsp. (15 mL) of liquid remains.

In a medium bowl combine the mustard, cherries, port and vinegar. Gradually whisk in the olive oil. Season with salt and pepper. Pour the vinaigrette over the greens and toss well. Divide between the plates.

Broil the baguette slices for 1 minute, or until just bubbling. Place one crouton on each salad. Place a piece of prosciutto at an angle to the crouton and serve.

LESLEY STOWE

Frisée Salad with Provençal Phyllo Sticks

Serves 6

I am more often drawn to first courses than I am to entrées. Innovative, tasty and small, they leave that "I'd like a little bit more" feeling. Ménage à Trois was one of the first restaurants to play on this theme. Located on Beauchamp Street in London, everything they served was like a starter, even the desserts. They were particularly clever at pairing contrasting textures, temperatures and flavours; hence this irresistible salad.

1/4 cup	red wine vinegar	60 mL
1 tsp.	Dijon mustard	5 mL
1 Tbsp.	tapenade	15 mL
1/2 tsp.	sea salt	2.5 mL
1/8 tsp.	freshly ground black pepper	.5 mL
3/4 cup	extra virgin olive oil	180 mL
4	heads baby frisée, or 1 head regular frisée	4
1 recipe	Provençal Phyllo Sticks	1 recipe

Whisk together the vinegar, mustard, tapenade, salt and pepper. In a steady stream, whisk in the olive oil. Set aside. Wash and dry the frisée. Break into bite-size pieces.

Toss the greens in the vinaigrette just before serving. Divide the dressed greens among 6 plates, and top each serving with 2 warm phyllo stick halves.

Provençal Phyllo Sticks

Makes 6 phyllo sticks

1 Tbsp.	extra virgin olive oil	15 mL
1/2	yellow bell pepper, minced	1/2
1	red bell pepper, minced	1
1/4	red onion, minced	1/4
1 Tbsp.	minced garlic	15 mL
1/4 cup	pitted chopped kalamata olives	60 mL
1/4 cup	chopped feta cheese	60 mL
1/4 cup	chopped basil	60 mL
6	sheets phyllo pastry (page 149)	6
3/4 cup	melted butter	180 mL

Preheat the oven to 350°F (175°C). Line a baking sheet with parchment paper.

Heat the olive oil in a frying pan over medium-low heat and sauté the bell peppers, onion and garlic until soft and translucent. Let cool and add the olives, feta and basil. Mix to combine.

Lay out one sheet of phyllo pastry on your work surface. Cut in half widthwise. Brush one of the halves with melted butter, place the other half on top, and again brush with butter. Spoon 1/6 of the vegetable mixture across the bottom edge of the phyllo, leaving a 1-inch (2.5-cm) margin on each side. Fold each edge in towards the middle, then, starting from the bottom, roll the phyllo away from you, enclosing the mixture and creating a stick, or log. Repeat this procedure until you have 6 phyllo sticks. Brush them with butter and place on the prepared baking sheet. Bake for 10-15 minutes, or until golden brown. Let sit for 5 minutes, then cut each log in half on a diagonal. Serve warm.

Butter Lettuce Salad with Focaccia Croutons and Sun-Dried Tomato Dressing

Serves 4

My friend Richard and I enjoyed this salad on a daily basis while working at a popular Mediterranean restaurant. Rick would simply toss the butter lettuce leaves and warm croutons with lots of dressing and we would eat straight from the bowl. If you think your guests would prefer their own plates, you can prepare this slightly fancier version using wedges of butter lettuce, cherry tomatoes, focaccia croutons and shaved Parmesan cheese. It makes a wonderful luncheon salad topped with grilled prawns.

1	head butter lettuce, washed and spun dry	1
1	5 x 6-inch (12.5 x 15-cm) piece of focaccia	1
2 tsp.	olive oil	10 mL
3 Tbsp.	chopped sun-dried tomato	45 mL
3 Tbsp.	hot water	45 mL
2 tsp.	Dijon mustard	10 mL
1/2 tsp.	tomato paste	2.5 mL
1	clove garlic	1
1/2 cup	chopped fresh basil leaves	120 mL
1 1/2 Tbsp.	white wine vinegar	22.5 mL
1/4 cup	olive oil	60 mL
6 Tbsp.	tomato juice or water	90 mL
	fine sea salt and freshly ground black pepper to taste	
8	cherry tomatoes, cut in half	8
4 oz.	Parmesan cheese, shaved	113 g

Trim the core of the lettuce and cut through the core to make 4 wedges. Store the lettuce in the fridge until ready to serve.

Preheat the oven to 400°F (200°C). Cut the focaccia into 24 1-inch (2.5-cm) cubes. Toss the croutons in the 2 tsp. (10 mL) olive oil. Spread the croutons out over a baking sheet, and bake for 7 minutes. Set the croutons aside to cool.

MARY MACKAY

Soak the sun-dried tomato in the hot water for 5 minutes. Place the sun-dried tomato and soaking water, mustard, tomato paste, garlic clove, basil leaves and vinegar in a blender. Purée on high speed for about 15 seconds. Use a spatula to scrape down the sides of the blender. Add the $1/4$ cup (60 mL) of olive oil and blend until the dressing is smooth. Add the tomato juice or water, salt and pepper, and blend on high speed until smooth. The dressing should be slightly thickened, but thin enough to pass through a squirt bottle. Add more tomato juice or water if the dressing is too thick. Transfer the dressing to a small squirt bottle. If you do not have a squirt bottle, you could place the dressing in a heavy plastic bag and make a small cut in the bottom corner when ready to squeeze out the dressing.

Squirt a dollop of dressing on 4 cold plates. Place a wedge of butter lettuce, cut side up, on each plate. Arrange the croutons and cherry tomatoes on each salad and squirt dressing over top of each salad. Top each serving with shaved Parmesan cheese and more freshly ground black pepper to taste.

tip: **Spice-Infused Oils**

Infused spice oils are quick and easy to make. This works well with curry, cumin or coriander, to name a few. Make sure your spices are full of flavour. Mix about 1 Tbsp. (15 mL) spice with enough water to loosen the mixture to the consistency of ketchup. Warm in a small saucepan until just fragrant, about 30 seconds. Mix with 1 cup (240 mL) lightly flavoured oil, such as canola, and pour into a clean jar. For the best flavour, let stand overnight or several hours. Keep refrigerated. Use to garnish plates, brush on breads for toasting or flavour a sauté or pasta. **–GM**

Wild Greens with "Potted" Cheddar and Fig Vinaigrette

Serves 6

This vinaigrette was originally inspired by a delicious commercial fig vinegar that was difficult to procure, so we decided to create our own. Drizzle it on grilled vegetables, splash it on braised greens, stir a little into your pan drippings from a roast—this vinegar adds a little touch of magic. The "potted" cheddar is tasty and will remind you of a good "cheese ball" served at parties in days gone by.

6 oz.	aged Cheddar cheese, finely grated	170 g
1 Tbsp.	minced shallot	15 mL
2 Tbsp.	white wine or vermouth	30 mL
2 tsp.	Dijon mustard	10 mL
1/4 tsp.	Worcestershire sauce	1.2 mL
1/4 tsp.	freshly ground black pepper	1.2 mL
2 tsp.	chopped fresh parsley	10 mL
3 Tbsp.	Fig Vinaigrette	45 mL
2 tsp.	honey	10 mL
	salt and freshly ground black pepper to taste	
1/3 cup	extra virgin olive oil	80 mL
8 cups	mixed wild greens	2 L

Combine the cheese, shallot, wine or vermouth, mustard, Worcestershire sauce, pepper and parsley in a medium bowl. Mash together with the back of a wooden spoon until well combined. With wet hands, shape the cheese into 6 little cakes. (You could also spread the "potted" cheddar onto crostini or small pieces of toast and serve these with the salad.)

Whisk the vinaigrette, honey, salt and pepper together in a small bowl. Slowly whisk in the olive oil.

Place the greens in a bowl and drizzle with the dressing. Toss gently. Divide the greens among 6 plates and top each with a cake of cheese.

MARGARET CHISHOLM

Fig Vinaigrette

Makes 2 cups (475 mL)

1/2 cup	dried black Mission figs, loosely packed	120 mL
2 cups	balsamic vinegar	475 mL
2 tsp.	vanilla extract	10 mL
3 Tbsp.	honey	45 mL

Place the figs and vinegar in a non-reactive (stainless steel, glass or enamel-lined) pot. Bring to a boil. Remove from heat, cover and let sit for 1 hour. Place in a blender and process until very smooth. Add the vanilla and honey and blend for a few seconds. Store in a glass jar or bottle in the fridge (it will keep for up to 6 months). Shake before using.

quick bite: **Parmesan Crisps**

Make sure you use a good-quality Parmigiano-Reggiano for these crisps. (Asiago also works well.) They make a great garnish for salads and appetizers. Preheat the oven to 350°F (175°C). Line a baking sheet with parchment paper. Spread coarsely grated Parmigiano-Reggiano in 2-inch (5-cm) circles. The cheese should be loosely spread with holes and gaps. Bake until the cheese starts to bubble and they are a light golden brown. Remove the pan from the oven and let them cool. They will crisp as they cool, and will keep easily for 3–4 days. I store them loosely wrapped in layers of paper towel in a container. –DC

MARGARET CHISHOLM

Fennel and Caramelized Pepper Pear Salad with Fig Vinaigrette

Serves 6

I assembled some favourite flavours, creating this salad for a cooking class benefitting our Les Dames d'Escoffier scholarship fund. On a rainy winter night the colours were warm and rich, the flavours sweet, yet spicy. After a full house and a fun class, we focussed on some serious product testing. Indulge, we did. Regrets, none! The cheese is available at specialty shops; you can substitute Cambozola.

1/2 cup	dried Mission figs, chopped	120 mL
1/2 cup	water	120 mL
1/2 cup	white wine	120 mL
1/2 cup	apple juice or cider	120 mL
1/4 cup	balsamic or sherry vinegar	60 mL
1/3 cup	olive oil	80 mL
1/4 tsp.	sea salt	1.2 mL
	freshly ground black pepper to taste	
3	Bosc or Bartlett pears, ripe but not soft	3
2 Tbsp.	unsalted butter	30 mL
2 Tbsp.	dark brown sugar	30 mL
1 tsp.	coarsely cracked black pepper	5 mL
1	large fennel bulb, trimmed (2 bulbs if small)	1
3 cups	cress or mixed seasonal salad greens	720 mL
2 Tbsp.	olive oil	30 mL
	salt and freshly ground black pepper to taste	
6 oz.	Gorgonzola mascarpone strata cheese	170 g
1/2 cup	walnut pieces, toasted	120 mL

Prepare the dressing ahead. In a non-reactive saucepan, combine the figs with the water, wine, juice or cider and vinegar. Bring to a boil, reduce the heat and cook until just soft. Transfer the stewed fig mixture into the food processor or blender and purée. Pour the purée into a bowl and whisk in the 1/3 cup (80 mL) olive oil. Season with salt and pepper.

Peel and core the pears and slice them into thick slices lengthwise. Melt the butter in a skillet large enough to hold the pears in a single layer. Add the pears and brown sugar. Cook until the sugar has melted and the pears are tender and a deep caramel colour. Add the cracked pepper and stir to coat. (Extra juice that develops around the pears will be added to the salad.) The peppered pears may also be made ahead; rewarm to serve.

Fennel darkens when cut, so preparing it at the last minute gives the freshest look and texture. (It won't be as crisp but it can be cut ahead of time, sprinkled with lemon juice and covered with plastic wrap.) The finer the slices, the better flavour the fennel will have. Use a plastic mandoline slicer to shave the bulb across into fine rings, or halve the bulb lengthwise, core and place cut side down. Slice lengthwise into the thinnest possible slices, using a chef's knife.

To assemble the salad, toss the greens with the 2 Tbsp. (30 mL) olive oil and season lightly with a pinch of salt and pepper. Divide the greens among 6 plates. Stack alternating layers of pears and fennel on the greens, drizzling with the pear juice from the pan. Top each salad with slices of Gorgonzola strata and walnut pieces. Just before serving, drizzle the fig vinaigrette and any remaining pear juice in concentric circles around the salad on the plate.

Caesar Salad Napoleons
with Parmesan Phyllo Crisps

Serves 6

Mary Mackay's love of bread salads—for obvious reasons—inspired me to look at an old favourite and have some fun. I also love the elegant flare of the Caesar salad at the French Laundry in Napa Valley, with its many components and attention to detail. This version is not nearly as daunting. If a little stacking isn't your style, then serve it simply, with the crisps on the side.

2	heads Romaine lettuce, washed and trimmed	2
2	eggs	2
6 Tbsp.	lemon juice	90 mL
	freshly ground black pepper to taste	
2	cloves garlic, minced	2
1	can anchovies, including oil	1
1/2 cup	grated Parmesan cheese	120 mL
1/2 cup	mildly flavoured olive oil	120 mL
1 recipe	Parmesan Phyllo Crisps	1 recipe
6 oz.	wedge Parmesan or Grana Padano cheese	170 g

Separate the small, tender leaves at the heart of the romaine and chill. Cut the outside leaves into a chiffonade of fine "shredded" strands. (If you do not like the crunch of the lower stem, slice it out before stacking or use the top section of the leaves.) Work in batches, stacking several leaves at a time. Cut across the stacked leaves, as thinly as possible. You should have about 3 cups (720 mL). Chill until needed.

To make the dressing, combine the eggs, juice, pepper, garlic, anchovies and grated Parmesan in a blender or food processor. Blend until a smooth paste forms. With the motor running, add the oil in a continuous thin stream, processing until the oil is incorporated and the mixture thickens. Taste and adjust if needed—it may need more lemon or oil to balance the flavours. Transfer to a clean container—a squeeze bottle with a nozzle is excellent—and chill until needed.

To assemble the salads, use the hearts of romaine as the base for stacking; combine with some of the dressing and toss to coat. Divide them between six serving plates, mounding them in the centre.

Top the base layer with a Parmesan phyllo crisp. The second layer is the chiffonade lettuce. Dress the lettuce lightly with some of the dressing and mound it on the crisp. Add some Parmesan shavings. For the third layer, add another crisp and a smaller amount of chiffonade. Drizzle with dressing and top with Parmesan shavings. Serve at once.

Parmesan Phyllo Crisps

Makes 18 pieces

3	sheets thawed phyllo pastry (page 149)	3
1/2 cup	unsalted butter, melted	120 mL
1/2 cup	grated Parmesan cheese	120 mL
1 Tbsp.	coarsely ground black pepper	15 mL

Preheat the oven to 350°F (175°C).

You will need two 11 x 17-inch (28 x 43-cm) baking sheets and two sheets of parchment to fit inside one baking sheet. Unroll the phyllo sheets. Remove 1 sheet and cover the remaining phyllo with a damp towel to prevent drying. Fit the phyllo in the baking sheet lined with parchment. Brush with melted butter and sprinkle generously with 1/3 of the cheese and pepper. Top with the second sheet of phyllo, pressing to seal. Brush the second sheet with butter and sprinkle with another 1/3 of the cheese and pepper. Repeat with the last layer.

Cover the phyllo with the second sheet of parchment. Place the second pan on top of the other to hold the sheets flat. Bake for 15 minutes, or until brown. When the sheets are baked, cut into pieces or simply break them into shards. Store in a cookie tin until needed. They will keep for several days if they're stored in a cool, dark place.

My late mother and grandmothers
started me off well —— from the
basics to picking fiddlehead ferns.

MY FIRST INSPIRATIONAL CHEF, outside of my family, was Graham Kerr, the Galloping Gourmet. That's where I first learned how to use a knife properly. I clearly remember one dish—squab with a sour cream and cherry sauce. To this, he added the bird's liver! I was simultaneously aghast and transfixed. My head spun with questions. Eating pigeons? Do you just go and get one from a roof? What would it taste like? Liver and cherries? From that moment on, I was hooked.

I had myself pegged for a pastry chef and my first restaurant job had me baking the "in" foods of the late '70s— carrot cakes, cheesecakes and quiches. The owner, Taro, and his partner, Tiger, were

Japanese, and they introduced me to the wonders of Japanese food. Of all cuisines, this is the one that I still have the most awe and respect for.

Ottawa gave me my second and third ethnic cuisine passions—Middle Eastern and Chinese food. I would study the shelves of grocery stores, trying to figure it all out and too shy to ask questions.

In Toronto, I was blessed to be able to work with Vanipha Southalack and her family. From her, I learned how to make curries, noodle soups, meat salads called *lap* and all the amazing nuances of this food.

Working at the David Wood Food Shop was about as close to heaven as imaginable at the time. David is still inspiring me

KAREN BARNABY

KAREN BARNABY

now with his Salt Spring Island goat and sheep milk cheeses.

This could not end without honouring my family and the huge role they have played. My late mother and grandmothers started me off well—from the basics to picking fiddlehead ferns. Visiting my aunts, uncles and cousins was a treat because there was always something new to discover.

Just before I started writing this, I was speaking to my wise friend Diane and bemoaning how my sources of inspiration seem to have dwindled in the last ten years. "Well," she said, "when you were younger, you absorbed things and took them inwards. Now you are the inspiration and are putting yourself outwards."

Thanks, Diane, that's a source of inspiration!

Warm Asparagus with
Cool Goat Cheese Sabayon

Serves 4 to 6

Asparagus and goat cheese are a natural combination. David Wood's truffled Salt Spring Island goat cheese makes an even better combination, but the effect can be simulated by scattering droplets of truffle oil over the dish. This sabayon is not difficult and it is a great sauce to have in your repertoire. Try it with salmon.

6	large egg yolks	6
1/2 cup	white wine	240 mL
6 oz.	soft unripened goat cheese, at room temperature	170 g
1 Tbsp.	milk	15 mL
1 Tbsp.	lemon juice	15 mL
1/2 cup	whipping cream	120 mL
2 Tbsp.	minced chives	30 mL
2 lbs.	fresh asparagus, woody stems snapped off and, if desired, peeled	900 g

Whisk the egg yolks and white wine together in a heatproof bowl. Place the bowl over, not in, a pot of simmering water. Continue whisking until the mixture becomes thick and triples in volume, about 4-5 minutes. Set the bowl into a larger bowl filled with ice water, or refrigerate until cool, whisking occasionally.

Beat the goat cheese until smooth. Gradually beat in the milk and lemon juice. Fold in the chilled egg yolk mixture. Beat the whipping cream until soft peaks form. Fold into the mixture. Stir in half the chives. Cover and refrigerate. Use the same day.

Tie the asparagus into two bundles with string. Bring a large pot of water to a boil. Salt liberally. Add the asparagus and cook until tender.

Serve the asparagus hot; or serve it cold, first cooling it under cold running water. Drain well, remove the string and arrange on a platter or individual plates. Top with the sauce and sprinkle with the remaining chives. Serve immediately.

K A R E N   B A R N A B Y

Carrot and Beet Salad
with Toasted Mediterranean Couscous

My former chef, Anne Milne, prepared one of my favourite salads—a healthy dish with shredded carrots, wild rice, currants and hazelnuts tossed in a citrus dressing. I have substituted toasted Mediterranean couscous for the wild rice, and added beets for flare. Mediterranean couscous is a larger, toasted variety of couscous that can be purchased in specialty food stores. The crispness of the shredded raw veggies give this salad a great texture, and it has an incredible colour palette of orange, purple and pink.

Serves 4 to 6

1/4 cup	toasted Mediterranean couscous	60 mL
3 Tbsp.	orange juice	45 mL
1 Tbsp.	lemon juice	15 mL
1 Tbsp.	lime juice	15 mL
1 Tbsp.	honey	15 mL
	zest of 1 small orange	
	fine sea salt and freshly ground black pepper to taste	
2	medium carrots, peeled and shredded	2
1	medium beet, peeled and shredded	1
1 cup	toasted chopped hazelnuts	240 mL

Bring a small pot of salted water to a boil over high heat. Add the couscous, reduce the heat to medium and simmer for 5 minutes. Cover with a lid and remove from the heat; let the couscous soak for 20 minutes. Drain and rinse the couscous under cold water. Set aside.

Whisk together the orange juice, lemon juice, lime juice, honey and orange zest. Season with salt and pepper.

Toss together the couscous, citrus dressing, carrots, beets and hazelnuts. Serve immediately.

MARY MACKAY

Roasted Beet and Vegetable Salad
with Parmesan Pepper Dressing

Serves 6 to 8

This salad was created out of the need for something fresh, healthy and absolutely delicious. The dressing, so I've been told, is out of this world. Feel free to use it on any salad or as a dipping sauce for fritters, chicken or baked potatoes.

1 lb.	small beets	455 g
2	red onions	2
1/3 cup	olive oil	80 mL
1 1/2 lbs.	fresh asparagus	680 g
1	red bell pepper, julienned	1
1	yellow bell pepper, julienned	1
1	head radicchio, chopped	1
2	carrots, julienned	2
12	red cherry tomatoes, halved	12
12	yellow cherry tomatoes, halved	12
6	radishes, thinly sliced	6
10	white button mushrooms, thinly sliced	10
1/2 cup	fresh flat-leaf parsley	120 mL
1 cup	sunflower sprouts	240 mL
1 recipe	Parmesan Pepper Dressing	1 recipe
1/2 cup	sunflower seeds	120 mL
1/3 cup	slivered almonds, toasted	80 mL
1/4 cup	sesame seeds, toasted lightly	60 mL

Preheat the oven to 350°F (175°C).

Peel and quarter the beets and onions. Place them on a cookie sheet. Drizzle with olive oil and roast for about 40 minutes, or until the beets are soft and the onions caramel in colour. Set aside to cool.

Blanch the asparagus in boiling water for about 2 minutes, then place in ice water to stop the cooking process. Drain and place on kitchen towels to absorb the remaining moisture. Set aside.

To assemble the salad, place all the roasted and fresh vegetables in a large glass bowl. Add the parsley and sprouts. Pour the dressing over and toss gently so as not to break the vegetables by over-mixing. Garnish the salad with the seeds and nuts.

Parmesan Pepper Dressing

Makes 1 cup (240 mL)

1/2 cup	buttermilk	120 mL
1/3 cup	good store-bought mayonnaise	80 mL
1/2 cup	freshly grated Parmesan cheese	120 mL
1	clove garlic, minced	1
1	shallot, finely chopped	1
1 tsp.	freshly cracked Tellicherry pepper	5 mL
	sea salt or fleur de sel to taste	

Whisk together all the ingredients until the dressing is smooth and creamy. Chill for at least 1 hour to set the flavours before serving.

tip: **Cracking Nuts**

I learned this trick for cracking nuts from the hazelnut man who frequents our local market every fall. Place the nut on a cutting board. Using a sturdy mug, give the nut a whack . . . it cracks perfectly every time and does not shatter. **–MC**

Roast Potato Salad with Pancetta

Serves 4 to 6

I like the idea of a lightly roasted potato salad. It has an interesting texture and the warmth of the potatoes allows the flavour of the vinaigrette to be absorbed. This is a great accompaniment to grilled meat, fish or chicken. When I serve this salad with chicken I like to toss in 2 Tbsp. (30 mL) of finely grated Parmesan at the end. You can also toss the potato salad with fresh arugula or baby spinach just before serving; the residual heat from the potatoes will lightly wilt the greens.

2 lbs.	Yukon gold potatoes, peeled and cut in 1/2-inch (1.2-cm) dice	900 g
3 Tbsp.	olive oil	45 mL
	salt and freshly ground black pepper to taste	
1 Tbsp.	olive oil	15 mL
3	large shallots, diced	3
3 oz.	pancetta, cut into julienne	85 g
1 tsp.	whole grain mustard	5 mL
1 Tbsp.	Dijon mustard	15 mL
1 tsp.	honey	5 mL
3 Tbsp.	white wine vinegar	45 mL
4 Tbsp.	chicken stock	60 mL
5 Tbsp.	olive oil	75 mL
2 Tbsp.	chopped fresh chives	30 mL
1 tsp.	fresh thyme leaves	5 mL

Preheat the oven to 375°F (190°C).

Toss the potatoes in a medium bowl with the 3 Tbsp. (45 mL) of oil, salt and pepper. Place the potatoes on a baking sheet in an even layer and bake for 20 minutes, stirring occasionally, until they are just cooked through and a light golden brown.

In the meantime heat the 1 Tbsp. (15 mL) of oil in a small sauté pan over medium heat. When the oil is hot, add the shallots and cook for 2–3 minutes. Add the pancetta and cook until the bacon is crisp, 2–3 minutes. Remove the bacon and shallots and set aside, leaving the oil in the pan.

In the same sauté pan, combine both mustards, the honey, vinegar and stock. Bring this mixture to a boil, remove the pan from the heat and slowly whisk in the 5 Tbsp. (75 mL) oil. Season to taste with salt and pepper.

Remove the potatoes from the oven and place them in a medium bowl. Add the pancetta-shallot mixture. Pour in the vinaigrette and toss. Add the chives and thyme, toss again and serve.

---

tip: **Roasting Garlic**

To roast garlic, preheat the oven to 350°F (175°C). Make a straight cut, about 1/4 inch (.6 cm) deep, across the stem end of the garlic bulb, exposing the cloves. Rub the bulb with olive oil and pour a little oil in the centre of a piece of foil. Place the garlic cut side down on the foil. Wrap the foil around the bulb. Roast for 1 hour. Don't be tempted to speed the process by raising the oven temperature—it will often result in bitter garlic. Different types of garlic from different growing regions take varying amounts of time, but it generally takes at least an hour for the sugar in the garlic to develop, giving it the characteristic mellow flavour. When it's done, it should be caramel-coloured and soft in texture. If not, return it to the oven for another 15 minutes. Let the garlic cool in the foil. Remove the cloves with a knife tip or squeeze them out of the papery husks.
–The Girls

Purple Peruvian Potato Salad

Serves 8

You love, I love, we all love potato salad. The question is: Do we need yet another potato salad recipe? The answer is *yes*, when it is purple and Peruvian. I discovered these potatoes in a market in Boston and searched them out in local markets. Save 3 or 4 and let them sprout, as I did. (Well, I cannot take credit; my husband, Jose, planted them and we had a delicious harvest of purple potatoes in the fall.)

3 lbs.	purple Peruvian potatoes	1.35 kg
3	bulbs garlic	3
2 cups	tiny frozen peas	475 mL
2 Tbsp.	Dijon mustard	30 mL
3 Tbsp.	tarragon vinegar	45 mL
2 Tbsp.	chopped fresh tarragon	30 mL
2 Tbsp.	chopped fresh chives	30 mL
1 Tbsp.	chopped fresh mint	15 mL
2/3 cup	roasted grapeseed oil	160 mL
	fleur de sel to taste	
	freshly ground Tellicherry pepper to taste	

Peel and quarter the potatoes. Place in a pot of lightly salted water and boil until just cooked through. Do not overcook them or they will fall apart. Drain, cool and set aside.

Cut about 1/4 inch (.6 cm) off the top of each garlic bulb. Rub the bulbs with a little olive oil, wrap in foil and bake for 45 minutes at 325°F (165°C). When cool enough to handle, separate and peel the cloves. Set aside.

Blanch the peas in boiling water for 2 minutes. Drain and place in ice water to stop the cooking process. Drain and set aside.

In a small mixing bowl, whisk together the mustard, vinegar, tarragon, chives and mint. Slowly whisk in the oil until the dressing is thick and smooth. Season with the fleur de sel and pepper.

To assemble the salad, place the potatoes in a shallow serving platter, pour the dressing over and toss lightly to mix. Do not over-toss or the potatoes will begin to break apart. Add the reserved peas and garlic cloves, and toss to coat. Refrigerate until serving time.

tip: **Label Your Party Platters**

When preparing for a buffet dinner, have all of your serving trays out and ready. I always label each dish with a small piece of paper indicating what it will hold. (Post-it notes work best, as they don't accidentally fall off.) When guests arrive and offer their help, gladly accept. You will never have food on the wrong plate! **–CMV**

CAREN MCSHERRY-VALAGAO

Warm Potato Salad

Serves 6

This is best served as an accompaniment to roasted or grilled meats. It could also be served on a bed of frisée or wild greens as an appetizer. Potatoes with a low starch content, such as baby red nuggets or new white potatoes, are the best choice.

2 Tbsp.	grainy mustard	30 mL
1 Tbsp.	sherry or red wine vinegar	15 mL
3 Tbsp.	olive oil	45 mL
	sea salt and freshly ground black pepper to taste	
6	slices crisp-cooked bacon	6
6 cups	new red or white potatoes, sliced $^1/_3$ inch (.9 cm) thick (about 1 $^3/_4$ lbs./800 g)	1.5 L
1 Tbsp.	unsalted butter	15 mL
1 $^1/_2$ cups	diced leek, white part only	360 mL
2 Tbsp.	chopped fresh parsley	30 mL

Whisk the mustard and vinegar together in a small bowl. Whisk in the olive oil. Season with salt and pepper. Chop the bacon into $^1/_2$-inch (1.2-cm) dice.

Cook the potatoes in salted water until tender but not water-logged. Drain and keep warm. While the potatoes are cooking, heat the butter in a sauté pan over low heat. Add the leek and sauté until tender. Set aside.

Combine the warm potatoes, bacon, leek, dressing and parsley. Serve warm.

Bocconcini with Kalonji Seeds, Cilantro and Tomato

This recipe combines my most recent love with my first–Indian and Middle Eastern food. There is a wonderful string cheese that you can find at Middle Eastern stores that is flavoured with kalonji seeds. Kalonji seeds are a frequent addition to many Indian dishes and are available at Southeast Asian markets. The seed comes from the flowering plant called love-in-a-mist and is sometimes mistakenly called black onion seed. One of my culinary heroes, Madhur Jaffery, includes a salad made with fresh cheese in one of her books. While it is illuminating to make the cheese, it is not always practical, so I tried it with bocconcini, adding a few flourishes of my own. It is absolutely fantastic with buffalo mozzarella and way more *cool* than bocconcini and tomato salad.

Serves 4

4	large bocconcini or 2 buffalo mozzarella	4
	sea salt and freshly ground black pepper to taste	
1/2 tsp.	kalonji seeds	2.5 mL
4 Tbsp.	finely diced red onion	60 mL
1/2 cup	seeded, finely diced tomato	120 mL
2 Tbsp.	coarsely chopped cilantro leaves	30 mL
2 Tbsp.	coarsely chopped fresh mint leaves	30 mL
2 Tbsp.	lime juice	30 mL
2 Tbsp.	lightly roasted pine nuts	30 mL
1 Tbsp.	roasted black sesame seeds	15 mL

Slice the cheese into 1/4-inch (.6-cm) slices and arrange them on a serving dish. You may slightly overlap them if you wish. Sprinkle with the salt, pepper and kalonji seeds.

Combine the onion, tomato, cilantro, mint and lime juice. Season lightly with salt. Spoon over the sliced cheese and sprinkle with the pine nuts and sesame seeds. Serve immediately.

KAREN BARNABY

Taco Salad

Serves 6

Think summer, think margaritas, think patio eating, think taco salad. This salad is an art form that gets its personality from inexpensive tortilla forms, available at most good cooking stores. The shells can be made in advance and stored for at least 2 to 3 days. Summer entertaining is not supposed to be rocket science, nor should it be a ball and chain. It is all about good food, fast, that tastes amazing.

To make the baskets:

6	8-inch (20-cm) flour tortillas	6
3 Tbsp.	olive oil	45 mL
1 tsp.	ground cumin	5 mL
1 tsp.	sea salt	5 mL

Preheat the oven to 350°F (175°C). Have ready the small-size tortilla forms or a 6-inch (15-cm) stainless steel bowl.

Mix the oil, cumin and salt together in a small bowl. Place a tortilla on your work surface and brush it with the oil mixture. Fit the tortilla inside the form or lay it over the outside of an inverted stainless steel bowl. Press so that the tortilla lies flat against the form. Bake for about 8 minutes, or until golden brown. Remove and cool. Repeat for the remaining tortillas.

For the salad:

1	head iceberg lettuce	1
2 cups	cooked black beans	475 mL
1/2	small sweet onion, sliced	1/2
1 cup	pitted kalamata olives	240 mL
1	large avocado, diced	1
4	ripe Roma tomatoes, diced	4
1	small jicama, peeled and diced	1
1 cup	grated Monterey Jack cheese	240 mL
1 cup	grated aged Cheddar cheese	240 mL
1 cup	your favourite salsa	240 mL
1/3 cup	olive oil	80 mL
1 cup	sour cream	240 mL

Finely shred the lettuce and place it in a large mixing bowl. Add the beans, onion, olives, avocado, tomatoes, jicama and both cheeses. Toss together to mix well. At this point the salad can be covered and chilled until serving time.

To serve, stir the salsa and olive oil together. Pour it over the salad and toss well to coat all the ingredients. Fill the prepared tortilla shells with the salad and garnish with a dollop of sour cream.

---

tip: **Slicing Basil**

Drizzle your basil leaves with olive oil before chopping. This will keep the basil from going black. Use a sharp knife and a slicing motion to avoid bruising the tender herb. Basil should never be finely chopped; just slice it into strips.
–MC

CAREN MCSHERRY-VALAGAO

Roast Garlic Flatbread
with Chicken and Greens

Serves 6

This recipe uses a pizza dough for the flatbread. You can grill or bake the chicken breasts. Use your favourite salad greens and any dressing that appeals. You can use any vinaigrette, or a creamy Parmesan dressing would work as well.

1 Tbsp.	dry yeast	15 mL
2 tsp.	honey	10 mL
1 cup	warm water	240 mL
3 cups	unbleached flour	720 mL
1 ^{1}/$_2$ tsp.	salt	7.5 mL
1 ^{1}/$_2$ tsp.	olive oil	7.5 mL
2 Tbsp.	olive oil	30 mL
4 Tbsp.	roasted garlic (about 2 bulbs)	60 mL
6 Tbsp.	grated Parmesan cheese	90 mL
	freshly ground black pepper to taste	
4	5-oz. (140-g) boneless skinless	4
	chicken breast halves	
1 lb.	salad greens or baby spinach	455 g
	dressing of your choice	

In a small bowl dissolve the yeast and honey in ^{1}/$_4$ cup (60 mL) of the warm water. Let sit until the mixture starts to bubble, 5–8 minutes.

In a food processor with a dough hook, combine the flour and salt. With the mixer on medium-low, add the 1 ^{1}/$_2$ tsp. (7.5 mL) olive oil and the yeast mixture. Add the remaining ^{3}/$_4$ cup (180 mL) of warm water and knead for 5 minutes, until the dough becomes smooth and elastic. Remove the dough.

If you don't have a machine with a dough hook, mix the dough by hand. Knead it for 10–15 minutes on a lightly floured board. Add more flour as you knead, but no more than is absolutely necessary.

Place the dough in a greased bowl, cover it with a damp towel and allow it to rest in a warm place for 30 minutes.

Flatten the dough and divide it into 6 equal pieces. Stretch and work each piece of dough for about 30 seconds and roll it tightly into a smooth ball. Cover the dough again with a damp towel and let rest for 20 more minutes. Place the balls of dough on a lightly floured surface and roll each into a 6-inch (15-cm) circle. If you are making the dough ahead, you can stack the individual pieces of rolled dough with waxed paper between each one. Cover snugly with plastic wrap and refrigerate.

Preheat the oven to 400°F (200°C).

Place the rounds on a lightly oiled baking sheet. Place them in the oven and bake for 5 minutes. Remove the partially baked flatbreads from the oven, brush the edges with a little of the 2 Tbsp. (30 mL) olive oil and cool them to room temperature. Divide the roasted garlic between the rounds. Using a spoon, spread a thin layer on each flatbread. Sprinkle with Parmesan cheese and pepper.

Heat the remaining olive oil in a medium sauté pan over medium-high heat. Season the chicken breasts with salt and pepper and sear each side for 2 minutes. Place the chicken in the oven to finish cooking, 8–10 minutes. While the chicken is cooking, finish the flatbreads. Bake until they are golden brown and the cheese has melted, about 3–4 minutes.

To serve, cut each flatbread into 4 wedges. Place 4 wedges on each plate. Toss the greens in the dressing and place on top of the flatbread triangles. Slice each chicken breast on the diagonal into 4 or 5 thin slices and divide the slices among the 4 plates, on top of the greens.

Warm Mahi Mahi Salad

Serves 6

When our children were small and we were not hostage to spring break, we spent a great deal of time on the island of Maui. Every Friday the local market would be our destination. The vendors arrived very early and so did we, in order to pick the best offering. Mahi mahi was always my favourite—this is one of the creations we enjoyed for lunch while basking in the sun sipping a crisp Sauvignon Blanc. Oh, those definitely were the days. Lulu's balsamic vinegar is available at specialty food stores. If you can't get it, substitute your favourite balsamic.

2	bulbs garlic	2
18	small silverskin onions	18
2 Tbsp.	olive oil	30 mL
2 lbs.	mahi mahi	900 g
3 cups	cooked small white navy beans	720 mL
3	large Roma tomatoes, diced	3
1/3 cup	virgin olive oil	80 mL
2 Tbsp.	Lulu's fig-infused balsamic vinegar	30 mL
1 Tbsp.	capers	15 mL
2 tsp.	lemon zest	10 mL
1/2 cup	chopped fresh parsley	120 mL
	sea salt and freshly ground black pepper to taste	
2 cups	fresh arugula, washed and dried	475 mL
1/2 tsp.	toasted chopped macadamia nuts	2.5 mL

To roast the garlic, cut 1/4 inch (.6 cm) off the top of each bulb and rub the whole bulb generously with olive oil. Wrap in foil and bake at 325°F (165°C) for about 45 minutes, or until soft. Cool, then separate and peel the cloves and set aside.

Blanch the onions in boiling water for about 3 minutes. Cool under cold running water. The skins will then slip off easily. Heat a cast-iron frying pan over medium heat and add the 2 Tbsp. (30 mL) of oil. Toss in the peeled onions and sauté until they begin to brown, shaking the pan every few minutes. Set aside.

Brush the fish with a little olive oil. Broil or pan-fry the fish until it is cooked through, about 6-8 minutes, or until it flakes easily. Set aside.

In a bowl, combine the garlic cloves, onions, beans, tomatoes, $^1/_3$ cup (80 mL) oil, balsamic vinegar, capers, lemon zest and parsley. Stir to combine and season with sea salt and pepper.

To serve, place a handful of arugula on a serving plate, spoon a generous portion of the bean mixture on top, and crown with the mahi mahi. Garnish with toasted macadamia nuts. Serve warm or cold.

---

tip: **Peeling Onions**

To minimize tears and aggravation when peeling and cutting onions, cut both ends off first. Then cut it in half, top to bottom, before peeling. The first layer peels off easily and the onion lies flat for chopping. **–MC**

CAREN MCSHERRY-VALAGAO

soups

Aunt Terry's Tomato Soup
with Margaret's Chèvre Crostini

Serves 4

I recently made a very simple tomato soup and the memory of Aunt Terry's kitchen came flooding back. She is my mother's youngest sister—well dressed, sophisticated—and we adored her. She taught us about fashion and other important matters. She also made great fudge. I've tried to capture the essence of her late summer soup. The goat cheese crostini will make you look more sophisticated.

2 Tbsp.	unsalted butter	30 mL
2 Tbsp.	finely diced onion	30 mL
1 tsp.	fresh thyme	5 mL
3 1/2 lbs.	ripe tomatoes, peeled, seeded and diced	1.6 kg
1 tsp.	sea salt	5 mL
	freshly ground black pepper to taste	
1/2 cup	whipping cream (optional)	120 mL
4 oz.	chèvre	113 g
1	small clove garlic, very finely chopped	1
1 Tbsp.	lightly chopped fresh basil	15 mL
4	slices baguette	4

Melt the butter over medium-low heat and add the onion. Cook and stir until soft and translucent. Add the thyme, tomatoes, salt and pepper. Simmer for 10-15 minutes. Purée in a blender or food processor. Return to the pot and add the cream, if desired. Heat until hot and keep warm.

Mash the chèvre with a spoon and mix in the garlic and basil.

Toast the baguette slices and spoon on the chèvre mixture. Place under the broiler and broil until just warm and lightly toasted at the edges. Float one crostini in each bowl of soup.

Carrot and Beet
Salad with Toasted
Mediterranean
Couscous
(page 63)

Caramelized Pear
and Prosciutto Pizza
(page 28)

Rolled and Stuffed
Chicken Breast
(page 99)

Indian Butter Prawns
(page 118)

Cream of Cauliflower Soup

This soup was inspired by the love of cauliflower, tarragon and simplicity. The béchamel sauce adds a different dimension. It is perfect for a week-night dinner, or is easily dressed up for Saturday night.

Serves 8 to 10

2 Tbsp.	olive oil	30 mL
1/2 cup	finely chopped onion	120 mL
1 cup	diced celery	240 mL
1	small carrot, grated	1
1	head fresh cauliflower, separated into florets, approximately 1 lb. (455 g)	1
8 cups	chicken stock	2 L
1	large bay leaf	1
1 tsp.	dried tarragon	5 mL
1/2 tsp.	freshly ground 5-pepper blend	2.5 mL
	sea salt to taste	
1/4 cup	unsalted butter	60 mL
2/3 cup	unbleached flour	160 mL
3 cups	2% milk	720 mL
1 tsp.	sea salt	5 mL
1/2 cup	finely chopped fresh parsley	120 mL

Heat the oil in a soup pot over low heat. Add the onion and let it sweat until it is transparent but not browned. Add the celery, carrot and cauliflower. Cover and cook on low for 10 minutes, stirring frequently.

Add the stock, bay leaf, tarragon and pepper. Season with salt. Bring to a low boil, then reduce and simmer for 15 minutes. Meanwhile, prepare the béchamel.

Melt the butter in a non-aluminum pot. Stir in the flour. Cook over low heat for about 2-3 minutes. Slowly whisk in the milk. Continue whisking until the sauce is smooth. Add the 1 tsp. (5 mL) salt.

Slowly stir the béchamel sauce into the soup. Add the parsley and cook for 10 minutes. Ladle into bowls and enjoy.

CAREN MCSHERRY-VALAGAO

Roasted Tomato Tequila Soup

Serves 6

Roasting the tomatoes adds a deep flavour dimension to this soup. I always use Roma tomatoes because they are meatier, with less liquid, and produce a more intense flavour and thicker soup. The tequila gives it the kick—leave it out if you prefer.

2	dried pasilla chili peppers	2
15	medium-size ripe Roma tomatoes	15
	sea salt and freshly ground black pepper to taste	
5	cloves garlic, chopped	5
1/2 tsp.	Mexican oregano	2.5 mL
5 Tbsp.	olive oil	75 mL
1	large white onion, diced	1
2 tsp.	smoked Spanish paprika	10 mL
6 cups	chicken stock	1.5 L
1/2 cup	chopped fresh cilantro	120 mL
1/2 cup	chopped fresh parsley	120 mL
1/3 cup	tequila	80 mL

Preheat the oven to 400°F (200°C). Cut the chilies open and place them in a hot, dry cast-iron pan over medium-high heat. Toast them for about 2 minutes on each side. Remove the stem, place the chilies in a bowl and cover with hot water. Let them soften for about 30 minutes. Strain off the water and set the peppers aside.

Cut the tomatoes in half and place them on a cookie sheet in a single layer, cut side up. Sprinkle with salt, pepper, garlic and oregano. Drizzle with 3 Tbsp. (45 mL) olive oil. Place in the oven and bake for about 1 hour. Cool.

Transfer the cooled tomatoes to the bowl of a food processor, along with the drained chilies. Purée and set aside.

Heat the remaining 2 Tbsp. (30 mL) oil in a soup pot over medium-high heat. Add the onion and sauté until nicely browned. Add the paprika. Cook and stir 1 more minute.

Add the chicken stock, puréed tomato/chili mixture, cilantro and parsley. Simmer on low until heated through. Adjust the seasoning to suit your taste. Ladle into bowls and top with a splash of tequila. Serve hot or cold.

quick bite: **Barbecued Duck, Tofu and Daikon Soup**

I make a quick zip through Chinatown on my way home to pick up the ingredients for this soup. I love the sweet and slightly turnip-y flavour of daikon. Try it in a beef stew.

Combine peeled and chopped daikon, a few slices of ginger and chicken stock in a pot. Bring to a boil and reduce to a simmer. Cook until the daikon is tender, about 15-20 minutes. Add one package of soft tofu, stirring vigorously to break up the tofu, and the meat from one Chinese barbecued duck. (You can use barbecued pork if you prefer.) I like to add a splash of fish sauce, a squeeze of lime and chopped green onion, cilantro and lots of chilies to my bowl. **–KB**

I've had many inspirations over the years.

SOME OF MY EARLIEST memories involve the smell of baking bread. In Newfoundland any wife and mother worth her salt baked her own—every day. My mother was the envy of the neighbours with her little family of only four children. In every house I visited as a child there was bread rising, bread being punched down, bread baking, bread being consumed in great slabs with partridgeberry jam. For those who couldn't wait, there were toutins. This was bread dough fried in butter or margarine, a wonderful treat. There was always a clamour for the bread straight from the oven, so there would be a tray of small buns in addition to the loaves. Other favourites were warm gingerbread with Nestlé's cream, rhubarb pies and apple pies, cookies and squares of all descriptions. I got in trouble for making mud pies with real eggs and milk, but as a result I got a Kenner Easy Bake Oven (remember those?). I would make cookies with sprinkles on top as bribes for my older brothers.

Fast-forward 20 years and 5,000 miles to the Comox Valley, British Columbia. My eyes were opened to the amazing quantity, freshness and variety of fruits and vegetables. I was inspired. I had to cook. I had to garden. The first cookbook I owned was *The Joy of Cooking.* From it I learned to make beef stew and chicken à la king, jams, jellies and homemade applesauce. There was nothing I wouldn't try. Home-canned

DEB CONNORS

salmon? Not a problem. With the kitchen windows steamed up and my second-hand pressure cooker rocking on the stove, I expected to die at any moment, but I would not be stopped. I had some triumphs and plenty of disasters, but I was hooked.

I've had many inspirations over the years. My friend Lorna Bridge, who owns Country Catering in Courtenay, B.C., was one of the first. She is an amazing cook who likes nothing better than to feed people. Ferdinand Bogner, chef at The Old House in Courtenay where I did my apprenticeship. His drive and professionalism were truly inspiring. The unflappable Andrew Howarth. My sister, Diane, who, when I visit her house in Seal Cove, is so easily impressed. My sous-chefs Mario and Sioux, who have amazing energy and are invaluable in the day-to-day running of Aqua Riva. And my general manager at the restaurant, Ann Bentley, who has a true passion for food and the industry and who has always challenged me to do my very best.

Fresh Tomato and Chorizo Soup with Chipotle Cilantro Cream

Serves 6 to 8

This soup was inspired by a walk through the Granville Island Market just down the street from where I live. The perfectly ripe tomatoes beckoned and the chorizo hollered as I walked past. This soup is best if it's made with ripe tomatoes. If they are un-available, use good-quality canned tomatoes, substituting 1/2 cup (120 mL) of drained canned tomatoes for each fresh tomato. The tomatoes set off the spicy garlic of the chorizo sausage. Chipotles—smoked jalapeño peppers—are readily available in very small tins; freeze the leftovers. If you choose, substitute a good quality, milder pasilla chili power.

3/4 lb.	chorizo	340 g
3 Tbsp.	olive oil	45 mL
6	slices bacon, chopped	6
1	medium white onion, diced	1
3	cloves garlic, minced	3
2	stalks celery, diced	2
8	ripe tomatoes, chopped	8
4 cups	chicken stock	950 mL
	salt and freshly ground black pepper to taste	
1 recipe	Chipotle Cilantro Cream	1 recipe

Preheat the oven to 350°F (175°C). Place the chorizo in a baking pan and cook for 15-20 minutes. Remove the sausages from the oven and place them on a paper towel to drain and cool. When the sausages are cool, cut them in 1/2-inch (1.2-cm) slices. Set aside.

Heat the olive oil in a large saucepan over medium-high heat. Sauté the bacon for 2-3 minutes. Add the onion, garlic and celery and sauté for 2 minutes. Reduce the heat to medium and cook for 5 minutes, until the onions are translucent. Add the tomatoes and cook for 5 minutes, then add the chicken stock. Season with salt and pepper. Bring the soup to a boil, reduce the heat to low and simmer for 30 minutes.

DEB CONNORS

Transfer the soup to a food processor or blender and purée it. Strain it through a sieve and then return it to the saucepan. Add the sliced chorizo and simmer the soup for 10 minutes. While the soup is simmering, make the Chipotle Cilantro Cream.

To serve, ladle the soup into soup bowls and place a dollop of the cream in the centre of each bowl.

Chipotle Cilantro Cream

Makes approximately $2/3$ cup (160 mL)

1/2 cup	sour cream	120 mL
1 tsp.	chipotle purée or chili powder	5 mL
1	small tomato, seeded and finely diced	1
1/2 tsp.	freshly ground black pepper	2.5 mL
1 tsp.	freshly squeezed lime juice	5 mL
2 tsp.	chopped fresh cilantro	10 mL

Combine all the ingredients in a small bowl. Refrigerate.

Curried Cauliflower Soup
with Apple and Almond Relish

Serves 6 to 8

Soups are often overlooked, I think, because they are perceived as labour intensive. Nothing could be further from the truth—this soup is fast and nutritious, and you can always use canned low-sodium chicken broth in place of the stock. The curry accents and the garnish take this soup out of the realm of the ordinary and give the cauliflower a new dimension.

3 Tbsp.	vegetable oil	45 mL
1	medium white onion, diced	1
2	stalks celery, diced	2
2	cloves garlic, chopped	2
3 Tbsp.	curry powder	45 mL
1/2 tsp.	ground cumin	2.5 mL
4 cups	cauliflower florets	950 mL
6 cups	chicken or vegetable stock	1.5 L
2 tsp.	freshly squeezed lemon juice	10 mL
2 tsp.	salt	10 mL
1 tsp.	ground white pepper	5 mL
1/2 tsp.	hot sauce	2.5 mL
1 tsp.	Worcestershire sauce	5 mL
1 cup	whipping cream	240 mL
1 recipe	Apple and Almond Relish	1 recipe
1 Tbsp.	finely chopped fresh chives	15 mL

Heat the vegetable oil in a large saucepan over medium heat. Add the onion, celery and garlic and sauté for 2 minutes. Add the curry powder and cumin and sauté for 3 minutes. Add the cauliflower and sauté for 2 minutes. Add the stock and lemon juice and bring the soup to a boil. Turn the heat to medium-low and simmer for 20 minutes.

Pour the soup into a blender or food processor and purée. Strain the soup and return it to the saucepan. Add salt, pepper, hot sauce and Worcestershire sauce. Heat to a simmer and whisk in the cream.

To serve, ladle the soup into shallow soup plates and place the relish in the centre of each serving. Sprinkle the chives over the soup.

Apple and Almond Relish

Makes approximately 3/4 cup (180 mL)

1 Tbsp.	butter	15 mL
1 tsp.	honey	5 mL
1	Granny Smith apple, peeled and finely diced	1
4 Tbsp.	coarsely chopped toasted almonds	60 mL
1 Tbsp.	finely chopped fresh chives	15 mL

Melt the butter and honey in a small sauté pan over medium heat. Add the apple and almonds and cook gently for 5 minutes. Remove from the heat and cool. When the relish has cooled to room temperature, stir in the chives.

tip: **Clarified Butter in the Microwave**

Cut the butter into 1-inch (2.5-cm) cubes and place them in a glass bowl. Cover the bowl tightly with plastic wrap and microwave on high for 5–6 minutes. Let the bowl cool a little before removing it from the microwave. Skim any foam from the top of the butter and gently pour or ladle the clarified butter into another container, leaving the residue in the bottom of the glass bowl. You can store the cooled butter in a tightly covered container in the refrigerator for 3–4 weeks. **–DC**

DEB CONNORS

Arugula Soup with Rosemary Garlic Oil

Serves 6

My grandmother Alvina's visits were always greatly anticipated. Granny would come once or twice a year for a few weeks to help my mother, who was raising eight children. Granny would cook and bake up a storm. The big batches of molasses cookies and shoe boxes full of date squares were very popular with us kids. Then there were the weed soups and boiled beef tongue sandwiches for lunch. I now love wild greens soup, but I still don't like tongue. Italian grannies make a similar soup with arugula. The best arugula I ever tasted was at an open-air market in the little town of Alba, where it grows wild just like my granny's dandelions or lamb's quarters.

2 Tbsp.	butter	30 mL
1/2 cup	diced onion	120 mL
1	clove garlic, minced	1
5 cups	water or chicken stock	1.2 L
1 tsp.	fresh thyme leaves	5 mL
2	large potatoes, cut into 1/2-inch (1.2-cm) dice	2
	salt to taste	
1/4 tsp.	freshly ground black pepper	1.2 mL
4 cups	arugula or lamb's quarters, chopped into 1-inch (2.5-cm) pieces	950 mL
1/2 cup	freshly grated Parmesan cheese	120 mL
3 Tbsp.	Rosemary Garlic Oil (optional)	45 mL

Heat a medium saucepan over medium-low heat. Add the butter, onion and garlic. Cook until soft and tender, about 10 minutes. Add the water or stock, thyme and potatoes. Simmer for 20–30 minutes, or until the potatoes are very tender. Season with salt and pepper. Add the arugula or lamb's quarters and simmer for 2 minutes.

MARGARET CHISHOLM

Serve in warm bowls with a spoonful of Parmesan cheese and a drizzle of Rosemary Garlic Oil, if desired.

Rosemary Garlic Oil

Makes $^{1}/_{2}$ cup (120 mL)

1/2 cup	extra virgin olive oil	120 mL
6	cloves garlic, roughly chopped	6
3	3-inch (7.5-cm) sprigs rosemary	3

Heat everything together in a small pot over low heat until the garlic just begins to turn pale gold, approximately 3 minutes. Make sure not to overheat it or the oil will lose its fragrance.

Remove from the heat. Allow to sit for 20 minutes or longer. Strain and discard the rosemary and garlic. This oil will keep for a couple of weeks in the refrigerator. It adds a warm herbal note to soups and can be drizzled over fish or crostini.

Bistro-Style Roasted Onion Soup

Serves 8

Many years ago, in southwestern France, I spent many hours in a bistro with walls and floors of stone and huge wood-burning ovens. They would mix pounds of onions, leeks, shallots and even the wild ramp for the next day's soup and let them slowly melt in a warm oven while we slept. During the day, their dogs would sleep on the floor among the table legs, chasing rabbits in their dreams, while the soup was served. I also dream—about the soup—and this is my version.

1/2 cup	finely diced double-smoked bacon	120 mL
2 Tbsp.	olive oil	30 mL
6	shallots, peeled and sliced	6
2	leeks, white and pale green part, thinly sliced	2
1	red onion, sliced	1
1	medium white onion, sliced	1
1	yellow onion, sliced	1
1 tsp.	sugar	5 mL
1/2 cup	white wine	120 mL
1 Tbsp.	fresh thyme leaves	15 mL
1/2 tsp.	salt	2.5 mL
	freshly ground black pepper to taste	
4-6 cups	chicken or beef stock	950 mL-1.5 L
	shaved Parmesan cheese	

Preheat the oven to 425°F (220°C).

Combine the bacon and oil in the bottom of a non-reactive roasting pan or large ovenproof skillet. Add all the shallots, leeks and onions, tossing to mix. Sprinkle with the sugar. Roast the onions, stirring them occasionally, until they're nicely caramelized. The total roasting time may vary from 45–60 minutes. The onions don't have to be completely cooked, but they should be nicely coloured. This step may be done ahead.

GLENYS MORGAN

If your roasting pan is suitable, use it to finish the soup, or transfer the onions to a soup pot. Any pan brownings left in the roaster should be deglazed and added to the soup pot. Simply heat the pan gently on the stove or in the oven, add some of the wine or stock to the hot pan, and stir to lift.

Add the wine, thyme, salt, pepper and just enough stock to cover the onions. (A good technique when making vegetable soup: keep the ratio of liquid to vegetables "tight" while the soup cooks. The flavour will be more intense and seasonings more easily adjusted.) Bring to a boil, then reduce the heat and simmer for 30 minutes to blend the flavours. The remaining stock can be added later to adjust and thin the consistency.

As a simple bistro soup, serve it very hot, adjusting the texture with more stock. Garnish each bowl with some shaved Parmesan, and serve with a good loaf of bread. For a thicker soup, purée about 1 cup (240 mL) of the soup and add it back in for body. For a more refined soup—or to extend leftover soup—purée the whole batch and thin with more stock or whipping cream. For the best flavour, make the soup a day or two ahead, allowing time for the flavours to marry.

Mom's Clam Chowder

Serves 6 to 8

My mother prepares her clam chowder following a recipe from Diane Clement's *Chef on the Run* cookbook. Mom adds tomatoes to hers and makes extra bacon crisps to sprinkle on top. I have added some fennel, capers and chili sauce to spike up mom's version. The soup is very chunky, and it's rich enough to serve as a main course with bread and a simple salad. If you want to dress up the soup for a dinner party, replace the canned clams with fresh steamed clams and garnish with crisps of sliced pancetta.

8	slices bacon, diced	8
1	small onion, diced	1
2	stalks celery, diced	2
1	small fennel bulb, trimmed of stalk and diced	1
3	medium Yukon gold potatoes, peeled and diced	3
1	clove garlic, minced	1
2 cups	vegetable stock or water	475 mL
1	14-oz. (398-mL) can diced tomatoes with juice	1
1	bay leaf	1
2 tsp.	capers	10 mL
2	5-oz. (140-g) cans whole baby clams with juice	2
1/2 tsp.	Chinese chili sauce	2.5 mL
1/4 cup	chopped fresh parsley	60 mL
1 1/2 cups	light cream	360 mL
	fine sea salt and freshly ground black pepper to taste	

In a large saucepan over medium heat, sauté the bacon until crisp. Use a slotted spoon to remove half of the bacon crisps. Wrap them in paper towel and set aside.

Add the onion, celery, fennel, potatoes and garlic to the saucepan and cook, stirring often, for about 10 minutes, or until the vegetables begin to soften.

Stir in the stock or water, tomatoes, bay leaf and capers and cook for 20 minutes, or until the potatoes are soft but not falling apart.

Add the clams and chili sauce and bring to a boil. Remove the saucepan from the heat and stir in the parsley and cream. Season with salt and pepper. Serve the soup in warm bowls with the reserved bacon crisps sprinkled over top.

tip: **Skinning Hazelnuts**

This is an odd method for removing the skins from larger quantities of roasted hazelnuts. A few years back we were using a lot of hazelnuts at the restaurant and my kitchen staff were moaning about the time it took to clean them. Someone had a eureka moment—we put the roasted hazelnuts in a loose cotton bag, tied the top tightly with string and bunged the bag in the dryer for 5 minutes. The motion of the dryer was enough to loosen the skins without breaking up the hazelnuts and the skins clung to the cotton. This method works very well, but you must be sure to tie the bag very tightly. **–DC**

Velouté of Artichokes
with Toasted Hazelnuts

Serves 6

My launch into the culinary world started at the apron strings of Chef Chambrette in Paris, a master of soups extraordinaire. This is one of my favourites that he passed on to me.

10	fresh artichokes	10
7 Tbsp.	butter	105 mL
7 Tbsp.	flour	105 mL
12 1/2 cups	chicken stock	3 L
1/3 cup	toasted hazelnuts, coarsely ground	80 mL
1 1/2 tsp.	sea salt	7.5 mL
4	egg yolks	4
1 1/4 cups	whipping cream	300 mL
	salt and freshly ground black pepper to taste	
	chopped chives	

Bring a medium saucepan of water to a simmer. Trim all of the leaves from the artichokes to reveal just the base. Plunge the artichoke bases into the simmering water for 2 minutes. Drain, dry and slice the artichokes very thinly, setting aside two of the sliced artichokes for later use. Sauté the remaining artichokes in 1 Tbsp. (15 mL) of the butter, and set aside.

In a large saucepan, melt the remaining 6 Tbsp. (90 mL) butter and add the flour. Cook over medium-high heat until just bubbling. Stir continuously to be sure it doesn't burn. Add the stock, hazelnuts, salt and the sautéed artichokes. Reduce the heat and simmer for 15 minutes.

Place the mixture in a food processor and blend. Pass the mixture through a fine sieve and return it to the pot. Whisk the egg yolks and cream until they are just combined. Stir into the soup. Add the reserved artichoke slices and return to the heat. Heat until it just comes to a boil. Season with salt and pepper. Garnish each serving with chives.

entrées

Maple-Glazed Country Ham Loaf

Serves 8

Homey, comforting, easy to make—this was a dish my mother made that I loved. I've tweaked it a bit and, as Emeril would say, kicked it up a notch. But my siblings would agree it is still extremely reminiscent of a staple dish from our childhood. Put a chilled slice between 2 pieces of great crusty white bread, add spicy fruit chutney, balsamic onions or just Dijon mustard, and you'll have the most satisfying sandwich you've had in ages.

1 lb.	ground ham	455 g
1/2 lb.	ground pork	227 g
1/4 cup	oatmeal	60 mL
1/2 cup	milk	120 mL
2 tsp.	Dijon mustard	10 mL
2	eggs	2
1/2 tsp.	sea salt	2.5 mL
1/4 tsp.	freshly ground black pepper	1.2 mL
1/4 cup	maple syrup	60 mL
1/3 cup	golden brown sugar	80 mL
1/4 cup	balsamic vinegar	60 mL
2 Tbsp.	Dijon mustard	30 mL

Preheat the oven to 450°F (230°C). Combine the ham, pork, oatmeal, milk, 2 tsp. (10 mL) mustard, eggs, salt and pepper, mixing well. Press the mixture into a greased loaf pan. Bake for 15 minutes. Turn the heat down to 350°F (175°C). Mix together the remaining ingredients and pour over the loaf. Continue baking for 1 hour.

Rolled and Stuffed Chicken Breast

Not chicken again! That all-too-familiar phrase indicates boredom with dinner. However, chicken is an economical, healthy alternative to red meat and a choice children often prefer. This dish brings together an interesting combination of flavours that will never be labelled boring. Look for tapenade in specialty food markets.

Serves 6 to 8

3	large chicken breasts, removed from the bird whole and deboned	3
2	Japanese eggplants	2
8 oz.	feta cheese	227 g
1/4 cup	chopped fresh mint	60 mL
1/2 cup	chopped fresh parsley	120 mL
1 cup	tapenade	240 mL
12	slices prosciutto di Parma	12
1	14-oz. (398-mL) jar roasted red peppers, drained	1
	freshly ground black pepper to taste	

Place the chicken between 2 sheets of waxed paper and pound with a heavy object like a frying pan. Give the chicken 2–3 smacks until it almost doubles in size. Slice the eggplant in 1/4-inch-thick (.6-cm) rounds. Brush with olive oil and pan-fry or grill over medium-high heat on both sides until golden brown. Mix together the feta, mint, parsley and tapenade.

To assemble, lay the chicken breasts skin side down on your work surface. Place the eggplant evenly over each breast in a single layer. Top each with 4 slices of prosciutto, then a few slices of red pepper. Top with 1/3 of the feta mixture. Roll each portion of filling between your palms, forming a sausage-like roll. Place a roll in the centre of each chicken breast. Give each a grinding of pepper. Roll up the chicken jelly-roll fashion and tie with string. Brush with olive oil. Barbecue over medium heat for 20 minutes, or until cooked through.

Cut the chicken into slices and serve hot or cold.

CAREN MCSHERRY-VALAGAO

Portuguese-Style Chicken
with Olive Oil Mashed Potatoes

Serves 3 to 4

I used to love going into Rebelo's, a Portuguese store in Kensington Market. They had a large take-out counter of prepared meat, sausages and fish. One of the highlights was tuna cooked with Portuguese red pepper paste. Being a fan of all things spicy, I started cooking chicken slathered with the paste. I first saw this unusual way of cutting chicken at a barbecue stand in Mexico. It seemed a little strange until I thought about it. The thighs, which are the thickest part of the chicken, are closest to the heat, while the breast is slightly curved away from the heat. The thickness of the chicken is more even, so all the parts cook at the same time. This works extremely well when barbecuing on a grill.

1	chicken, preferably free-range, 2–2 1/2 lbs. (.9–1.1 kg)	1
1/2 cup	Portuguese hot red pepper paste (look for Vinga or Melo's brand)	120 mL
2	cloves garlic, minced	2
2 Tbsp.	extra virgin olive oil	30 mL
1 recipe	Olive Oil Mashed Potatoes	1 recipe

Split the chicken through the breast (that's right, the breast). Flip it skin side up and hit it with the heel of your hand to flatten it. Combine the red pepper paste, garlic and oil. Slather over the inside and outside of the chicken. If you have the time, refrigerate the chicken for a few hours or up to overnight. If you don't have time, don't worry—the chicken will still be delicious.

Preheat the oven to 375°F (190°C). Place the chicken skin side up in a pan and cook for 1 1/2 hours. Let the chicken rest for 10 minutes. Transfer to a cutting board and cut the chicken in half, along the back. Cut into leg, thigh, wing and breast sections. Pour the pan juices over the chicken and serve immediately with the potatoes.

Olive Oil Mashed Potatoes

Serves 4

1 1/2 lbs.	russet potatoes	680 g
1/3 cup	extra virgin olive oil	80 mL
1/2–3/4 cup	whole milk, heated	120-180 mL
2 Tbsp.	unsalted butter	30 mL
	sea salt to taste	

Peel the potatoes and cut them into 1 1/2-inch (3.7-cm) pieces. Place in a deep pot and cover with cold water. Add enough salt to make the water taste like sea water and bring to a boil. Cook until the potatoes are tender, approximately 20 minutes.

Drain the potatoes well. If you have a ricer, press the potatoes through the ricer back into the pot. If not, return them to the pot and mash by hand until smooth. With a wooden spoon, slowly beat in the olive oil. Add the warmed milk a little at a time, until the potatoes are light and fluffy. You may not need all the milk. Stir in the butter and adjust the seasoning with salt if needed.

---

quick bite: **Salsa Inspirations**

With all the lively salsas in specialty stores, let them be the inspiration for a quick meal. Spoon some salsa—chipotle with tomato or peach with red pepper or corn with chiles—over some chicken breasts. Add a little olive oil and, if desired, a spoonful of Dijon mustard to thicken. Marinate the chicken breasts for a few minutes and they're ready to grill, bake or sauté! For fish, make a simple vinaigrette, whisk in some salsa and drizzle over your catch of the day, cooked as you like it. –GM

Lemon Herb Marinated Chicken Breasts

Serves 4

The bright flavours of this zesty chicken will make it a hit. I love it with roast potatoes or a creamy risotto. You can use almost any combination of fresh herbs. I leave the wing bone on the chicken for presentation, and the skin crisps up beautifully.

For the marinade:

1/4 cup	olive oil	60 mL
1/2 cup	canola oil	120 mL
3 Tbsp.	freshly squeezed lemon juice	45 mL
1 tsp.	finely chopped lemon zest	5 mL
1 Tbsp.	honey	15 mL
1 Tbsp.	chopped fresh oregano	15 mL
1 Tbsp.	chopped fresh thyme	15 mL
1 Tbsp.	chopped fresh parsley	15 mL
1 Tbsp.	chopped fresh rosemary	15 mL
1 Tbsp.	chopped fresh basil	15 mL

Combine all the ingredients in a medium bowl.

To prepare the dish:

4	7- to 8-oz. (200- to 227-g) chicken breasts, skin and wing bone left on	4
3 Tbsp.	olive oil	45 mL
	salt and freshly ground black pepper to taste	
1 cup	chicken stock	240 mL
1 tsp.	lemon juice	5 mL
2 tsp.	butter	10 mL

Place the chicken breasts in a non-reactive shallow pan and pour the marinade over top. Refrigerate and let the chicken marinate for a minimum of 4 hours and up to overnight.

Preheat the oven to 450°F (230°C).

Heat the oil in a heavy, ovenproof sauté pan over medium-high heat. Remove the chicken from the marinade and season with salt and pepper. When the oil is hot, sear the chicken breasts skin side down until they are nicely coloured, 3–4 minutes. Turn the chicken breasts over, and place the sauté pan in the oven. Bake for 20–25 minutes, until the juices run clear.

Remove the chicken breasts to a serving platter and pour off the oil from the pan. Place the pan over medium-high heat. Add the chicken stock and lemon juice and bring it to a boil. Use a wooden spoon to scrape the bottom of the pan. Cook until the stock is reduced by half. Add salt and pepper to taste. Remove from the heat, swirl in the butter and pour the reduced sauce over the chicken breasts.

tip: **It Burns My Buns!**

Even small variations in oven temperatures can cause baked goods to vary drastically. To avoid disappointment, purchase a good oven thermometer and keep it hanging on a rack in the oven at all times. My oven at home runs at least 25°F (10°C) hotter than what appears on the dial, so I adjust accordingly. –MM

DEB CONNORS

My first recollection
of inspiration was when
I was twelve.

MY FIRST RECOLLECTION OF INSPIRATION was when I was twelve and read of a thirteen-year-old boy whipping up an amazing meal for his family. Not to be outdone, I searched through my mother's cook-books, locked the family out of the kitchen and told them to come back at dinnertime. Unbeknownst to me they invited friends over and dressed in black tie for the occasion. The Apple Charlotte with Cream was the highlight, in my opinion, already being partial to desserts at that early age. I wouldn't make Apple Charlotte again for nine years, until I was twenty-one and travelled to France to learn from some very experienced, star-studded chefs at La Varenne Ecole de Cuisine in Paris. From my experience as a teenager, poring over *Mastering the Art of French Cooking* by Julia Child and Simone Beck while babysitting, to being the only female in the kitchen of a two-star restaurant in Paris, I have been inspired by numerous sources and people. Tasting, reading and talking about ideas, we continue to shape our styles of cooking.

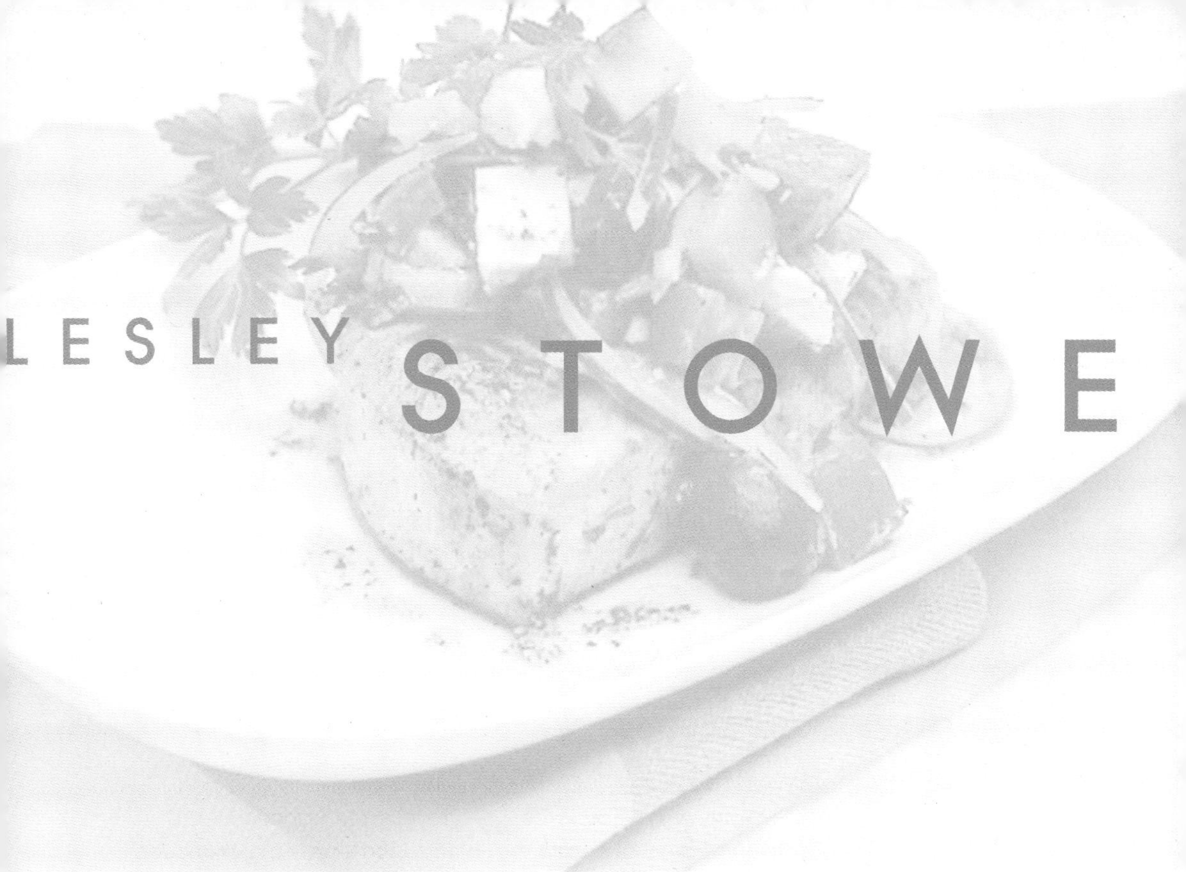

LESLEY STOWE

Chicken Augustina

Serves 6

My original recipe called for veal and was developed when a close friend got into the veal business and was trying to promote pink-veal to the local restaurants. I've substituted free-range chicken: delicious, more readily available, yet interesting enough to serve at a dinner party.

12	small shallots, peeled	12
2 Tbsp.	extra virgin olive oil	30 mL
	sea salt and freshly ground black pepper to taste	
2 cups	chicken stock	475 mL
3 Tbsp.	unsalted butter	45 mL
6	skinless, boneless chicken breast halves, preferably free-range	6
1/4 cup	dry white wine	60 mL
8	oil-packed sun-dried tomatoes, drained and cut into julienne strips	8
2 Tbsp.	cold unsalted butter	30 mL
1 Tbsp.	chopped fresh basil	15 mL
6 Tbsp.	toasted pine nuts	90 mL

Preheat the oven to 350°F (175°C). Coat the shallots with olive oil, season with salt and pepper and place in a shallow baking pan in the oven. Roast the shallots until brown and soft, about 30 minutes.

Place the chicken stock in a small pot over medium heat and cook until it's reduced to 1/2 cup (120 mL).

Heat the 3 Tbsp. (45 mL) butter in a skillet over medium-high heat and sauté the chicken for about 2 minutes per side. Transfer the breasts to a baking dish and bake in the oven for 15–18 minutes, or until firm to the touch. Remove from the oven and keep warm on a separate plate.

Pour off the butter from the skillet. Place the shallots, wine and sun-dried tomatoes in the skillet and cook over medium heat until the wine is nearly evaporated. Add the reduced stock and simmer until the sauce coats the back of a spoon. It should have the consistency of maple syrup. Whisk in the 2 Tbsp. (30 mL) cold butter, 1 Tbsp. (15 mL) at a time. Spoon the sauce over the chicken and sprinkle the basil and pine nuts on top.

quick bite: **Quick Quesadillas**

One of my favourite quick fixes is a fast preparation of quesadillas. The fillings can be whatever you have in the refrigerator—leftover cooked chicken, a handful of shrimp and so on. The more little bits of cheese you have, the better it is.

Lay 6 8-inch (20-cm) flour tortillas flat on your work surface. Prepare 2 cups (475 mL) of grated cheese. Spread half of each tortilla with the grated cheese, divide 1 cup (240 mL) cooked black beans over top and add any other leftovers you want. Fold over the undressed side. Place in a non-stick frying pan and cook on both sides until golden brown, about 2 minutes for each side. Cut into wedges and serve with your favourite salsa. **–CMV**

Salmon à la Trudeau

Serves 4

There is a long tradition of naming dishes in honour of a person—Crêpes Suzette, Steak Diana, Veal Orloff and so on. During the early stages of writing this book, former Canadian prime minister Pierre Trudeau passed away. I was deeply moved and decided to honour him with a dish. My roots are in the Maritimes, and I grew up in Montreal. I started my cooking career in the Rockies, where I met Mr. Trudeau on a hiking trip, and I learned to cook salmon in a fishing camp on the coast of British Columbia. This is a very Canadian dish: East Coast potato cakes meet West Coast salmon in a cross-country apple broth.

2	medium Yukon gold potatoes, peeled and cut in 1-inch (2.5-cm) cubes	2
1 Tbsp.	unsalted butter	15 mL
1/3 cup	milk	80 mL
1/3 cup	horseradish, finely grated, or from a jar	80 mL
2 Tbsp.	chopped chives	30 mL
1/2 tsp.	salt	2.5 mL
1/3 cup	diced onion	80 mL
1	6-inch (15-cm) sprig fresh thyme	1
2 1/2 cups	fresh apple cider	600 mL
1/4 cup	apple cider vinegar	60 mL
1/2 cup	white wine	120 mL
1 Tbsp.	unsalted butter	15 mL
4	5-oz. (140-g) salmon fillets	5
	salt and freshly ground black pepper to taste	

Preheat the oven to 400°F (200°C).

Place the potatoes in a small saucepan and cover with water. Add a dash of salt and bring to a boil. Boil gently until the potatoes are very tender. Drain and mash or put through a ricer. Spoon out $1/4$ cup (60 mL) potatoes and set aside for the sauce. To the remaining potatoes, add the 1 Tbsp. (15 mL) butter, milk, horseradish, chives and salt. Let cool. Divide the mixture into four. Wet your hands and shape the mixture into cakes roughly the size and shape of the salmon fillets.

Combine the onion, thyme, fresh cider, cider vinegar, wine and the reserved $1/4$ cup (60 mL) of potatoes in a small saucepan. Bring to a simmer and boil gently until reduced to 1 cup (240 mL). Strain, pressing on the solids with a rubber spatula, and keep warm. (You may do this ahead and reheat at serving time.) Just before serving, swirl in the remaining 1 Tbsp. (15 mL) butter.

Season the salmon with salt and pepper. Place the potato cakes on top of the salmon. Place on a non-stick baking sheet. Bake for 12 to 15 minutes, or until the salmon is barely firm.

Place the salmon in warm individual pasta bowls and surround with the warm apple jus.

---

tip: **Dried Porcini Mushrooms**

The longer these mushrooms soak the better. If you know you are using them the next day, go ahead and soak them overnight. Sometimes there is a bit of grit at the bottom of your soaking dish. Simply lift out the mushrooms and carefully pour off the mushroom liquid, leaving behind and discarding about 1 tsp. (5 mL), which will be gritty. Always incorporate the soaking liquid into your dish. −MC

Miso-Glazed Sablefish with Waldorf Salad

Serves 4

Sablefish or black cod has been one of my favourite fish since moving to the West Coast. I was so impressed with it that I couldn't believe that it was not widely eaten except in the Asian community. Happily, that is slowly changing. I have come up with many different preparations for it, including an osso buco-style dish and curing it like corned beef. This particular preparation is Japanese-influenced and the Waldorf salad provides a perfect foil for the rich and succulent texture of the sablefish.

1/3 cup	white miso	80 mL
1/4 cup	sugar	60 mL
2 Tbsp.	sake	30 mL
3 Tbsp.	mirin	45 mL
4	6-oz. (170-g) sablefish fillets, skin on	4
2	egg yolks	2
2 tsp.	lemon juice	10 mL
1 tsp.	Japanese prepared mustard or 2 tsp. (10 mL) Dijon mustard	5 mL
1/2 cup	vegetable oil	120 mL
2 Tbsp.	white miso	30 mL
	sprinkle of white pepper	
1/4 tsp.	grated lemon peel	1.2 mL
2 cups	Fuji or other flavourful apple cut into 1/4-inch (.6-cm) dice	475 mL
1/4 cup	celery hearts cut into 1/4-inch (.6-cm) dice	60 mL
1/2 cup	peeled daikon radish cut into 1/4-inch (.6-cm) dice	120 mL
2	green onions, thinly sliced	2
2 Tbsp.	pine nuts, toasted	30 mL
2 tsp.	black sesame seeds, toasted	10 mL

In the top of a double boiler combine the $^1/_3$ cup (80 mL) miso, sugar, sake and mirin. Cook for $^1/_2$ hour, stirring frequently. Remove from the heat and let cool. Coat the sablefish with the miso mixture. Cover and refrigerate overnight or up to 3 days.

In a food processor, combine the egg yolks, lemon juice and mustard. With the motor running, add the oil in a slow steady stream. When the mayonnaise has emulsified, add the 2 Tbsp. (30 mL) miso, pepper and lemon peel. Pulse to combine. Cover and refrigerate until needed.

When you are ready to serve the cod, combine the apple, celery, daikon, green onion, pine nuts and sesame seeds. Add the mayonnaise. Mix well and refrigerate the salad until ready to serve.

Position the oven rack approximately 8 inches (10 cm) under the broiler. Preheat the broiler on high. Place the fish skin side up on a baking tray and broil until browned and crisp around the edges, about 5 minutes. Turn over and broil 5 minutes longer. Serve immediately with the Waldorf salad on the side.

tip: **Designer Dogs**

Forget hot dog buns; serve hamburgers and hot dogs on croissants—it is much more sophisticated! –MM

Pan-Seared Swordfish
with Mediterranean Relish

Serves 4

This meaty fish is great for sautéing or grilling. Its mild flavour is nicely complemented by the savoury relish. You can also toss some of the relish with salad greens and serve the fish over top with the rest of the relish as garnish. The relish can be made earlier in the day, refrigerated and brought up to room temperature before serving. If you do this, add the feta cheese and avocado just before serving so that they do not break down.

For the relish:

1/2	avocado, cut into 1/2-inch (1.2-cm) dice	1/2
3 Tbsp.	extra virgin olive oil	45 mL
1 1/2 Tbsp.	freshly squeezed lemon juice	22.5 mL
2 Tbsp.	English cucumber, seeded and cut into 1/4-inch (.6-cm) dice	30 mL
2 Tbsp.	red bell pepper, cut into 1/4-inch (.6-cm) dice	30 mL
2 Tbsp.	yellow bell pepper, cut into 1/4-inch (.6-cm) dice	30 mL
3 Tbsp.	crumbled feta cheese	45 mL
1	Roma tomato, cut into 1/4-inch (.6-cm) dice	30 mL
1/4	red onion, cut in a fine julienne	1/4
2 Tbsp.	sliced pitted black olives	30 mL
1 Tbsp.	chopped Italian parsley	15 mL
1 tsp.	salt	5 mL
2 tsp.	freshly ground black pepper	10 mL

Place the avocado in a medium bowl. Pour in the oil and lemon juice. Add the cucumber, red and yellow peppers, feta cheese, tomato, onion, olives and parsley. Gently combine the ingredients and season with salt and pepper.

For the marinade:

1/4 cup	canola oil	60 mL
1	clove garlic, minced	1
1 Tbsp.	finely chopped fresh basil	15 mL
2 Tbsp.	white wine	30 mL
1 Tbsp.	freshly squeezed lemon juice	15 mL
1/2 tsp.	freshly ground black pepper	2.5 mL

Combine marinade ingredients in a small bowl and mix well.

To prepare the dish:

4	8-oz. (227-g) swordfish steaks, 3/4 inch (2 cm) thick	4
3 Tbsp.	canola oil	45 mL

Place the swordfish in a shallow glass dish. Pour the marinade over the fish, turning the steaks several times so they are well coated. Refrigerate and marinate for 2–4 hours.

Remove the fish from the marinade. Heat the oil in a large sauté pan over medium-high heat. When the oil is hot, sauté the swordfish for 3 minutes on one side. Turn the heat to medium and sauté the other side for 2 minutes more.

To serve, place each portion of swordfish in the centre of a large plate. Spoon the relish over the swordfish and serve immediately.

Black Cod with Braised Fennel and Blood Orange Sauce

Serves 4

Lavender is widely used in Provence to flavour foods—everything from fish to tapenade. Use it very carefully or it will taste like granny's soap. Perhaps each mouthful of food should have but one tiny speck of lavender on it. Adding lavender is optional but fun to try. In any case the whole combination of fish, fennel and orange works like a dream.

4	5-oz. (140-g) fillets fresh black cod (sablefish)	4
3/4 tsp.	sea salt	4 mL
	pinch freshly ground black pepper	
1 Tbsp.	extra virgin olive oil	15 mL
1/8 tsp.	lavender (optional)	.5 mL
1/2 tsp.	very finely chopped blood orange zest	2.5 mL
1	fennel bulb	1
1 Tbsp.	unsalted butter	15 mL
2 Tbsp.	water	30 mL
2 Tbsp.	sliced fresh basil	30 mL
1/2 cup	blood orange juice	120 mL
2 tsp.	raspberry or red wine vinegar	10 mL
2 Tbsp.	unsalted butter	30 mL
4	whole leaves fresh basil	4

Sprinkle the cod with salt. Dust with a tiny pinch of pepper. Drizzle with the olive oil and sprinkle with lavender, if desired, and blood orange zest. Let sit for 1 hour.

Preheat the oven to 350°F (175°C).

Trim the top stalks off the fennel. Trim the base and cut in half. Remove and discard the core and dice the bulb into 1/2-inch (1.2-cm) pieces. Warm a small saucepan over medium heat, add the 1 Tbsp. (15 mL) butter and the fennel. Cook and stir for 2 minutes. Add the water and a bit of salt and pepper. Cover and turn the heat down to low. Cook for 10 minutes, or until tender. If it gets dry, add a splash of water. Stir in the basil and keep warm.

Combine the blood orange juice and vinegar in a small saucepan. Place over medium heat and boil gently until reduced by half. Whisk in the 2 Tbsp. (30 mL) butter and a little salt and pepper. Keep warm.

Roast the cod in the preheated oven for 10-12 minutes, or until cooked through. It should be barely firm in the centre and flaking at the edges.

To serve, arrange a mound of fennel in the centre of a warm plate, top with a piece of cod and drizzle the sauce around the outside. Garnish with a leaf of basil.

quick bite: **Five-Minute Pasta**

Heat 2 Tbsp. (30 mL) of olive oil over medium heat. Add 2-3 cloves minced garlic and 1-2 Tbsp. (15-30 mL) chili flakes. Cook for 3-4 minutes. Toss in 4 cups (950 mL) cooked pasta and add another 2 Tbsp. (30 mL) of oil. Stir to coat the pasta. Sprinkle with $1/2$ cup (120 mL) Parmesan cheese and finish with fleur de sel and Tellicherry pepper to taste. Cook until the pasta is heated through. This recipe will serve 4 to 6 garlic and chili lovers. **–CMV**

Smoky Tomato "Paella" Risotto
with Basil and Garlicky Prawns

Serves 4

When Caren McSherry-Valagao first introduced me to her imported fire-roasted paprika, to me it was heaven in a can. The smell of a wood-burning fire, the colour of flames: it was a perfect match for tomato, prawns, rice—for a dish like paella! Seared scallops, steamed mussels, clams or spicy sausage would be delicious in this dish. For a more Spanish touch, replace the basil with roasted red and green bell peppers, peeled and diced.

16	prawns	16
4 Tbsp.	extra virgin olive oil	60 mL
3	ripe tomatoes, peeled, or canned Italian plum tomatoes, diced	3
	pinch sea salt	
4 cups	chicken stock	950 mL
2 Tbsp.	unsalted butter	30 mL
2	minced shallots	2
2 tsp.	fire-roasted Spanish paprika	10 mL
1 cup	Arborio rice	240 mL
1/4 cup	dry white wine	60 mL
1/2 tsp.	sea salt	2.5 mL
	freshly ground black pepper to taste	
2	cloves garlic, sliced	2
1 tsp.	lemon zest	5 mL
1/2 cup	fresh basil leaves	120 mL
	Parmesan cheese (optional)	

Peel the prawns, leaving the tail section intact. Score the prawns down the back and butterfly them. Toss with 1 Tbsp. (15 mL) of the olive oil. Refrigerate until needed.

To make the tomato base, heat 1 Tbsp. (15 mL) of the olive oil in a small saucepan. Add the tomatoes, cooking gently to heighten the flavour and soften, about 2–3 minutes. Season with a pinch of salt. This step may be done ahead.

Place the stock in a saucepan; heat to nearly boiling, then reduce to a simmer. Melt the butter and 1 Tbsp. (15 mL) of the olive oil in a Dutch oven over medium-high heat. For risotto, the pan should not be tall and narrow (stacking the rice) or too wide and shallow (liquid evaporates rapidly). Add the shallots and sauté, lowering the heat slightly, until soft and translucent. Lower the heat again, and add the paprika, being careful not to burn it. Add the rice, stirring to coat.

Raise the heat slightly and when the pan is just sizzling, add the wine. The wine will deglaze the pan and loosen the rice. Add enough hot stock to cover the rice and loosen it to a soupy consistency, then bring to a high simmer. When the rice is bubbling, add the tomato base. Add a ladleful of hot stock and stir until the stock is absorbed. Adjust the heat to a simmer. Season with salt and pepper.

Add enough stock to just cover the rice—some grains should be peeking through the surface. The rice should never drown or dry out. Add stock as needed, stirring well after each addition in a figure 8 motion, keeping the rice moving evenly around the pot. Taste frequently for an al dente texture. For most risotto recipes the ultimate consistency is creamy; it should be loose but not soupy (although some traditionally are) and not stodgy like cooled porridge. It is a balance between perfect consistency and texture!

When nearly ready, heat the remaining 1 Tbsp. (15 mL) olive oil over medium-low heat and gently cook the garlic without browning it. Remove the garlic slices and raise the temperature of the pan to high. Sear the prawns until they are just done and they've curled. Season with salt and pepper.

Remove the risotto from the heat and stir in the lemon zest. Taste for salt and pepper. Set aside a few basil leaves for garnish. Cut the remaining basil into a fine chiffonade: stack several leaves at once, roll gently like a cigar, then slice across the roll, creating very fine slices. Fluff the basil chiffonade and fold it into the risotto.

Spoon the risotto into pasta bowls. Top with the hot seared prawns, tails pointing up. Garnish with the fresh basil leaves. Serve at once, passing the Parmesan grater, if desired.

Indian Butter Prawns

Serves 4

I adore Indian butter curries. I have never had two that taste alike. When researching this recipe I asked several Indian ladies how they prepare their butter curries and each one gave me a different answer. Some add garlic, some do not, and some add diced tomatoes or tomato paste. What was common was the use of lots of butter and finishing the sauce with whipping cream to give the dish the richness I love. Butter curry is usually prepared with tandoori baked chicken, which can result in a rather lengthy recipe. Since prawns cook so quickly, they make a great substitution. The prawns are cooked in a generous amount of sauce, so you will need to serve this dish with steamed rice or Indian bread to mop up every last bit.

1/3 cup	plain yogurt	80 mL
1/2 tsp.	fine sea salt	2.5 mL
28	prawns, peeled and deveined	28
20	whole almonds, toasted	20
2	cloves garlic	2
2 Tbsp.	minced fresh ginger	30 mL
1 tsp.	curry powder	5 mL
1/2 tsp.	garam masala	2.5 mL
1/4 tsp.	chili powder	1.2 mL
2	green cardamom pods, ground	2
4 Tbsp.	unsalted butter	60 mL
5 Tbsp.	tomato paste	75 mL
1 cup	chicken stock, vegetable stock or water	240 mL
1 tsp.	honey	5 mL
1/3 cup	whipping cream	80 mL
1/3 cup	chopped fresh cilantro leaves	80 mL
	sea salt and freshly ground black pepper to taste	

In a medium bowl, stir together the yogurt and salt. Fold in the prawns and set aside while preparing the sauce.

Place the almonds, garlic, ginger, curry powder, garam masala, chili powder and cardamom in the bowl of a food processor. Use the steel knife attachment to blend the ingredients together.

In a large saucepan, melt the butter over medium heat. Add the almond and spice mix; cook, stirring often, for 3 minutes. Stir in the tomato paste and cook for another minute. Add the stock or water and honey and bring to a boil. Reduce the heat to medium-low and simmer the curry for 10 minutes. Pass the sauce through a fine mesh sieve, pressing down on the solids to extract all the liquid. Return the strained sauce to the saucepan.

Increase the heat to medium and stir in the prawns and cream. Cook until the prawns are cooked through, about 3-5 minutes. Stir in the cilantro and season with salt and pepper. Transfer the prawns and sauce to a warm dish and serve with steamed basmati rice or hot Indian bread.

tip: **Browning Butter**

Melt butter and let the milk solids in the bottom of the pan brown slightly. It gives butter a delicious nutty flavour that adds another dimension to waffle/pancake/crêpe batters. **–LS**

MARY MACKAY

Poor Man's Scallops with Penne

Serves 4 to 6

The first time I saw a whole halibut it was an 80-lb. (36-kg) specimen that had been hauled up onto the dock at a sports-fishing camp on the B.C. coast. Halibut cheeks from a fish this size would be as big as grapefruits. The texture and flavour of halibut cheeks is reminiscent of scallops, so if you can't find these delicacies locally, splurge and use scallops. Any smoked cheese will work, or use aged Cheddar and toss in a couple of strips of chopped crisp bacon at the end of cooking for the smoky flavour.

1 lb.	halibut cheeks	455 g
	sea salt and freshly ground black pepper to taste	
1 1/2 Tbsp.	coarse salt	22.5 mL
1 lb.	dried penne	455 g
2 Tbsp.	unsalted butter	30 mL
2 Tbsp.	olive oil	30 mL
2	cloves garlic, chopped	2
2	yellow bell peppers, cut into 1/2-inch (1.2-cm) dice	2
2	zucchini, cut into 1/2-inch (1.2-cm) dice	2
1 lb.	tomatoes, cut into 1/2-inch (1.2-cm) dice	455 g
1/2 cup	chopped fresh basil	120 mL
1 cup	grated smoked Cheddar cheese	240 mL
1/2 cup	freshly grated Parmesan cheese	120 mL

If the halibut cheeks are large, cut them into roughly 1-inch (2.5-cm) chunks. Season with salt and pepper.

Bring a large pot of water to a boil over high heat. Add the coarse salt and pasta to the boiling water. Cook, stirring occasionally, until the pasta is tender, but still firm to the bite, approximately 9–11 minutes. Drain the pasta and return to the pot.

While the pasta is cooking, heat a large skillet over medium-high heat. Add the butter and olive oil. When the foam subsides, add the halibut cheeks. Sauté for 2 minutes, turn over and sauté 2-3 minutes longer, or until barely cooked through. Remove the fish from the pan, set aside and cover to keep warm. Turn the heat down to medium, and add the garlic, bell peppers and zucchini to the pan. Cook 3-4 minutes. Add the tomatoes and basil. Cook and stir 1-2 minutes longer. Season to taste with salt and pepper.

Combine the fish, tomato mixture and pasta, toss gently and adjust the seasoning to taste. Transfer to a heated serving dish, add the cheeses and toss gently.

quick bite: **Quick Bean Gratin**

Keep a few cans of good-quality beans—garbanzo, lima, black—in the cupboard for emergencies, which includes having no time to cook them from scratch. I make a creamy garlicky gratin of white kidney beans when I need the taste of Italy in a hurry. Drain the beans and pour into a shallow baking dish. Add enough whipping cream to loosen, a few chopped garlic cloves, and if you have some, minced rosemary. Bake in a hot oven until bubbly. If you have any leftover, purée and enjoy on crostini. **–GM**

MARGARET CHISHOLM

Saffron Risotto with Shellfish and Prawns

Serves 4

This is a dish inspired by my friend Ann. She remembers every good meal she's ever had, describes them in great detail and then asks me to re-create them. This dish evolved from one of those descriptions. The creamy, rich risotto pairs beautifully with the lightness of the shellfish. The risotto sounds complicated, but it goes together quite quickly and is well worth the effort. You may not need all of the stock for the risotto; taste as you go.

For the risotto:

4 cups	chicken or vegetable stock	950 mL
2 tsp.	saffron threads	10 mL
4 Tbsp.	olive oil	60 mL
1/2 cup	minced white onion	120 mL
1 cup	Arborio rice	240 mL
	salt and freshly ground black pepper to taste	
2 Tbsp.	butter	30 mL
1/2 cup	grated Parmesan cheese	120 mL
1/2 cup	heavy cream, whipped	120 mL

Bring the stock and saffron to a boil in a saucepan. Reduce to low and keep warm.

Heat the oil in a large saucepan over medium heat. When the oil is hot, add the onion. Cook for 3–4 minutes, stirring with a wooden spoon, until the onion is cooked through but not coloured.

Add the rice and cook, stirring often, for 3–4 minutes, until the grains are somewhat transparent. Add $1/2$ cup (120 mL) of the hot stock and stir until the liquid is absorbed. Lightly season with salt and pepper. Continue adding the stock $1/2$ cup (120 mL) at a time, cooking and stirring until the liquid has been absorbed before adding more stock. The rice should continue to simmer throughout the cooking process. Taste the risotto to check the texture. When the rice is cooked, the grains should be plump and just tender, not mushy.

Remove the risotto from the heat and quickly mix in the butter, using the wooden spoon. Mix in the cheese and the whipped cream. Set aside and keep warm.

For the shellfish:

2 Tbsp.	olive oil	30 mL
24	mussels, debearded and scrubbed	24
16	clams, scrubbed	16
1	clove garlic, minced	1
$1/2$ cup	dry white wine	120 mL
2 tsp.	freshly squeezed lemon juice	10 mL
	salt and freshly ground black pepper to taste	
16	large prawns, shelled and deveined	16

Heat the oil in a large sauté pan over medium-high heat. Add the mussels, clams and garlic and sauté for 2 minutes. Add the wine, lemon juice, salt and pepper. Cover the pan with a lid and cook for 5–6 minutes, or until the shells have all opened. Remove the lid, add the prawns and cook for another 2–3 minutes, until the prawns are just cooked through. Discard any shellfish that haven't opened.

To serve, divide the risotto equally among 4 large shallow bowls. Scatter 6 mussels, 4 clams and 4 prawns over the risotto in each bowl. Spoon a little of the pan juices from the shellfish over each dish.

Cast-Iron Mussels with Smoky Aïoli

Serves 4

Sizzling hot fajita griddles are used to cook these mussels at Rose Pistola in San Francisco. I fell in love with the way the mussels plump in their own juices, making them perfect for dipping in a luscious sauce. They serve drawn butter, but I serve them with this Smoky Aïoli and crispy french fry potatoes for a relaxed dinner.

4 lbs.	mussels, scrubbed and beards removed	1.8 kg
1 tsp.	kosher or sea salt	5 mL
1 tsp.	cracked black pepper	5 mL
1 recipe	Smoky Aïoli	1 recipe
	lemon juice	

Heat a cast-iron skillet over high heat until it's very hot. Test the heat by pouring a spoonful of water into the pan; it should form beads that dance across the hot surface.

Spread the mussels in the pan in a single layer. The mussels will open and their juices will sear in the bottom of the pan. When they're wide open and the meat is plumped, they're ready to serve. Season generously with salt and pepper and serve from the pan. Drizzle with the Smoky Aïoli or pass bowls of aïoli for dipping. Add a squeeze of lemon to brighten.

Smoky Aïoli

Makes 1 cup (240 mL)

2	large egg yolks	2
1 tsp.	fire-roasted Spanish paprika	5 mL
2 tsp.	minced garlic	10 mL
1/2 tsp.	salt	2.5 mL
1 Tbsp.	sherry vinegar or white wine vinegar	15 mL
1/2–3/4 cup	extra virgin olive oil	120–180 mL

As with any hand-crafted mayonnaise, the flavour is best when it's made a few hours in advance. Whisk together the egg yolks, paprika, garlic, salt and vinegar. Slowly begin the emulsification by whisking the oil in drop by drop. When the mixture thickens like a pourable mayonnaise, add the balance of the oil in a steady stream, whisking constantly. The more oil whisked in, the thicker the aïoli. Taste for salt. Refrigerate for up to 4 days, but don't serve it cold; chilling dulls the olive oil flavour.

tip: **Keeping Mashed Potatoes Hot**

You can keep mashed potatoes hot by using a double boiler. Transfer the hot mashed potatoes to a non-metal bowl and place over a pot of simmering, not boiling, water. Place a piece of plastic wrap directly on the surface of the potatoes. The potatoes will keep hot for up to 2 hours.
–KB

GLENYS MORGAN

Beef Tenderloin with Portobello Boursin Sauce

Serves 8

One of my pilgrimages to San Francisco found a version of this dish in a tiny eclectic spot called the Flying Saucer. It may seem like a strange combination, but once you've tasted it you'll be sold.

3–4 lbs.	beef tenderloin	1.35–1.8 kg
	salt and freshly ground black pepper to taste	
6 Tbsp.	olive oil	90 mL
1 lb.	portobello mushrooms, cut in half and sliced	455 g
1 cup	sliced shallots	240 mL
4	cloves garlic, sliced	4
1/2 cup	coarsely chopped parsley	120 mL
4	sprigs thyme	4
1/2	sprig rosemary	1/2
8	peppercorns	8
1	small bay leaf	1
1 cup	dry white vermouth	240 mL
4 cups	dark veal stock	950 mL
6 oz.	black pepper Boursin cheese	170 g

Preheat the oven to 375°F (190°C).

Season the tenderloin with salt and pepper. Heat 3 Tbsp. (45 mL) of the olive oil in a heavy skillet over medium-high heat. Brown the tenderloin on all sides. Transfer to a roasting pan.

Pour the fat from the skillet. Add 2 Tbsp. (30 mL) of the olive oil to the skillet and sauté the mushrooms over medium-high heat until soft and golden. Season with salt and pepper. Remove from the pan and set aside.

Add the remaining 1 Tbsp. (15 mL) olive oil to the pan and stir in the shallots, garlic, parsley, thyme, rosemary, peppercorns and bay leaf. Sauté until soft, approximately 3–4 minutes. Add the vermouth and veal stock. Simmer until the liquid is reduced by $1/2$. Strain the sauce.

Place the beef in the oven and cook for 15–20 minutes. Remove and let rest. Return the sauce to the heat and reduce by $1/3$. Whisk in the Boursin and add the mushrooms.

Slice the beef tenderloin and arrange on plates. Spoon the sauce over and serve immediately.

tip: **Roasting**

Any time you are roasting anything, the key is to give the food space. The right-sized pan will allow hot air to circulate around the food; this takes away excess moisture and allows the food to caramelize or brown. A roast or chicken should have 2-3 inches (5-7.5 cm) on all sides; vegetables should be scattered in one layer over the pan. –MC

LESLEY STOWE

Vietnamese-Style Beef Brisket

Serves 4 to 6

There was a certain Vietnamese restaurant in Toronto that I used to frequent at least once a week for my beef brisket fix. It was mildly sweet and sour with a rich texture. After much contemplation on the subject, I figured out that the secret ingredient was . . . ketchup! Those were my more snobbish days and I could not admit that ketchup was the secret. Now, I gleefully don't care! Serve over fresh rice noodles, with bean sprouts, cilantro and green onions and a squeeze of lime, or with rice or French bread. This is a fabulous winter dish.

1-1 1/2 lbs.	beef brisket	455–680 g
1 Tbsp.	vegetable oil	15 mL
1 Tbsp.	finely chopped garlic	15 mL
2	1/2-inch (1.2-cm) slices fresh ginger, lightly crushed	2
2	stalks lemon grass, trimmed, cut into 3-inch (7.5-cm) pieces and lightly crushed	2
2	star anise	2
8 cups	water or beef stock	2 L
1 cup	ketchup	240 mL
1/2 tsp.	salt	2.5 mL
2	medium carrots, peeled and cut into 2-inch (5-cm) pieces	2
2	medium new potatoes, peeled and cut into quarters	2
1 Tbsp.	fish sauce	15 mL
1 Tbsp.	oyster sauce	15 mL

Cut the brisket into 1-inch (2.5-cm) pieces. Place in a pot, cover with cold water and bring to a boil. Simmer for 5 minutes and drain. Rinse under cold water and pat dry.

KAREN BARNABY

Heat the oil in a large pot over medium heat. Brown the beef in batches and remove to a plate. Add the garlic and ginger to the pot and cook, stirring, until the garlic turns a pale gold. Add the beef, lemon grass, star anise, water or stock, ketchup and salt. Bring to a boil, then reduce to a low, low simmer. Cover with a lid and cook for 2 1/2 hours, until tender, adding water only if the liquid falls below the level of the beef. The stew should be slightly thickened by this point. If not, cook over medium heat, uncovered, until it thickens.

Increase the heat to medium. Add the carrots, potatoes, fish sauce and oyster sauce. Cook, uncovered, for half an hour longer, or until the vegetables are tender. Taste and adjust the seasoning with salt and fish sauce. Like all stews, this is better if allowed to sit overnight.

Frikadeller

Makes 16 to 20 meatballs

At a young age my friend Anne introduced me to these tasty Danish meatballs. Her mother, Solvejg, usually had a few tucked away in the fridge for snacking. Frikadeller have a softer texture than Italian meatballs; they are very similar to miniature meat loaf. Serve them with potato salad or with sautéed red cabbage and new potatoes. My favourite way to eat frikadeller is in a meatball sandwich on Danish rye bread with remoulade (Danish mayonnaise mixed with relish), dried fried onions and lingonberry sauce. The condiments can be purchased in Danish delicatessens.

1 tsp.	unsalted butter	5 mL
1 cup	finely chopped onion	240 mL
6 oz.	ground pork	170 g
6 oz.	lean ground beef	170 g
6 oz.	ground veal	170 g
2 Tbsp.	unbleached all-purpose flour	30 mL
1 tsp.	fine sea salt	5 mL
1/2 tsp.	freshly ground black pepper	2.5 mL
1	large egg, beaten	1
1/2 cup	whole milk	120 mL
1 Tbsp.	unsalted butter	15 mL
1 Tbsp.	vegetable oil	15 mL

In a medium non-stick frying pan, heat the 1 tsp. (5 mL) butter over medium heat; cook the onion, stirring occasionally, for 5 minutes, or until softened. Allow to cool completely.

In a large bowl, mix together the cooked onions, pork, beef, veal, flour, salt, pepper, egg and milk. Cover the bowl with plastic wrap and refrigerate for 20–30 minutes.

Preheat the oven to 350°F (175°C). Line a baking sheet with aluminum foil.

Heat the 1 Tbsp. (15 mL) butter and oil in a large, wide frying pan over medium-high heat. Using a heaping tablespoon (15 mL) to measure each frikadeller, form into oblong balls, approximately 3 x 2 inches (7.5 x 5 cm). Brown 8 to 10 frikadeller at a time for 1 1/2 minutes on each side. Using a slotted spoon, transfer the browned frikadeller to the baking sheet. Continue to brown the remaining frikadeller, adding more oil to the pan if necessary.

Bake the frikadeller for 15–20 minutes, until cooked through. Drain on paper towel to absorb any extra grease. Serve hot or cold.

Two Fat Ladies Meatloaf

Serves 10 to 12

I adapted this meatloaf from the *Two Fat Ladies* cooking show. They called it hedgehog, probably because of its shape. Meatloaf makes a great breakfast, dinner or cold snack. Fry it and serve with eggs for breakfast. Or heat it up under the broiler, spread it with mayonnaise and Dijon mustard, then smother it with cheese and a few slices of tomato and broil until the cheese is bubbly. This makes a very pleasantly seasoned meatloaf with a flavour that is reminiscent of a pâté. Confirmed liver haters can omit the chicken livers.

1/2 lb.	mushrooms, diced	227 g
2 Tbsp.	unsalted butter	30 mL
	sea salt and freshly ground black pepper to taste	
1/2 lb.	chicken livers	227 g
1 1/2 lbs.	ground beef	680 g
1 1/2 lbs.	ground pork	680 g
1 lb.	sausage meat	455 g
1/2 cup	finely diced onion	120 mL
3	cloves garlic, minced	3
2 tsp.	ground coriander seeds	10 mL
1 tsp.	ground allspice	5 mL
1/4 tsp.	nutmeg	1.2 mL
2 tsp.	chopped fresh thyme or rosemary leaves	10 mL
2	eggs	2
1/2 lb.	sliced bacon	227 g
	bay leaves	
	branches of fresh rosemary	

Sauté the mushrooms in butter over high heat until the juices run, then season with salt and pepper. Set aside.

Remove the sinews from the livers. Lightly poach the livers in simmering water, just until they turn grey and firm up. Cool and dice. In a large bowl combine the liver, ground meats, sausage, onion, garlic, coriander, allspice, nutmeg and thyme or rosemary. Season with salt and pepper, keeping in mind that the sausage contains salt. Beat the eggs and add them to the mixture together with the mushrooms. Mix thoroughly with your hands.

Preheat the oven to 450°F (230°C). Oil a baking pan that has a rim and place the mixture in it, forming it into one or two oval shapes. (I like to make two smaller, torpedo-shaped loaves and reduce the cooking time by $1/2$ hour.) Cover with the bacon, criss-crossing the slices and tucking the ends under the meat loaf. Tuck some bay leaves and branches of rosemary into the bacon. Bake for 15 minutes, then lower the heat to 350°F (175°C) and cook for 1 $1/2$ hours for 1 large loaf or 1 hour for 2 smaller ones. Serve hot or cold.

Fiery Thai Beef with Herb Salad and Cooling Cucumber Pickle

Serves 4 to 6

Craving the flavours of an exotic cuisine—usually hot and spicy—is often my inspiration for dishes. I happily compromise on authenticity, creating my own arrangement of flavours, mixing techniques and styles. And it's got to be fast! Here it's a salad mix in the style of France with Thai herbs, a steak on the grill and a simple dressing whisked together with Thai staples. It's boldly flavoured, lean, healthy and goes especially well with an earthy Côtes du Rhône—I did say I was irreverent!

6 Tbsp.	kecap manis or dark soy sauce	90 mL
4	cloves garlic, crushed	4
3	limes	3
2-3	red bell peppers	2-3
2	flank steaks, approximately 1-1 $^1/_2$ lbs. (455–680 g)	2
1 cup	fresh basil leaves, stems removed	240 mL
2 cups	watercress, coarse stems removed, or mixed salad greens	475 mL
$^1/_2$ cup	mint leaves, stems removed	120 mL
$^1/_2$ cup	cilantro leaves with stem tops	120 mL
$^1/_2$ cup	Thai sweet chili sauce	120 mL
$^1/_4$ cup	soy sauce	60 mL
2	kaffir lime leaves	2
1 recipe	Cooling Cucumber Pickle (page 136)	1 recipe

Prepare the marinade for the beef and peppers by whisking together the kecap manis and garlic. Grate the zest from 2 of the limes before juicing them, then add zest and juice to the marinade.

Stem and core the bell peppers. Cut away any of the white membrane inside the peppers, then cut them lengthwise into fine strips, about $^1/_4$ inch (.6 cm) wide. These very fine strips soften as they cook and develop a nice glazed edge. (If grilling the peppers, leave them in larger strips and slice them after cooking.) Place the meat and peppers in a large baking dish or resealable plastic bag. Marinate for at least 1 hour at room temperature or up to a day refrigerated.

Combine the basil leaves (leave them whole or tear any large leaves) with the watercress. Mint is too pungent left in large pieces; stack the leaves, gently roll them like a cigar and then thinly slice across the roll, cutting a chiffonade of fine, thin strips. Add to the salad mix. For a subtle cilantro flavour, use only the leaves. For cilantro lovers, roll the leaves and tender stems into a loose ball and chop. Add to the salad mix. Refrigerate the salad greens until needed.

Prepare the salad dressing by whisking together the chili sauce and soy sauce. Zest the remaining lime using a citrus rasp or very fine grater. Add both the zest and the juice of the lime to the dressing. To mince the kaffir lime leaves, stack the leaves and slice in thin strips with a chef's knife. Mince the leaves as finely as possible and add them to the dressing. The dressing can be made up to 2 days ahead and refrigerated.

Let the steak return to room temperature before grilling. Remove the flank steaks from the marinade and pat dry with paper towels. Remove the peppers from the marinade with a slotted spoon. The marinade can be boiled to use as a basting sauce if desired.

Heat the grill, griddle or heavy skillet to very hot. Brush the steaks with oil and cook until nicely browned before turning and cooking the second side. Flank steak should be served pink in the centre for the best flavour and texture. Let rest, loosely covered, for about 10 minutes before carving.

(continued on following page)

GLENYS MORGAN

Sear the peppers on a grill until nicely browned on the edges. The peppers can be lightly seared and crisp or cooked until they are browned and soft; either texture is delicious. Remove from the grill and set aside. Carve the steaks across the grain and at an angle, making the pieces wide, but thin.

To assemble the salad, choose large dinner plates and divide the greens, mounding them in the centre of each plate. (A wide platter also makes a nice presentation.) Spoon a little of the dressing on the greens; arrange the slices of flank steak on top and finish with the peppers on top or around the plate. Spoon the remaining dressing around the plate for colour. Garnish with a mound of cucumber pickle.

Cooling Cucumber Pickle

Makes approximately 2 cups (475 mL)

1/2 cup	white wine vinegar	120 mL
1/2 cup	sugar	120 mL
1/2 cup	water	120 mL
1 tsp.	salt	5 mL
1	long English cucumber	1
1	red chili pepper, minced	1

Bring the vinegar, sugar, water and salt to a boil, cooking to dissolve the sugar. Cool the pickling brine completely before using. It can be made ahead and refrigerated until needed.

To decorate the cucumber skin, use a zester to scrape down the length of the cucumber, removing thin strips of the green skin. Cut the cucumber lengthwise into quarters and slice across as thinly as possible.

About 1 hour before serving, mix the brine with the cucumber and chili. Serve at room temperature or well chilled for contrast.

quick bite: **Tea and Tarragon Granita**

You only need a freezer and a spoon to make this Italian version of sorbet. I was searching for a new flavour of granita and the idea of tarragon and tea appealed to me. The tea proved to be very cleansing on the palate and the "high tones" of the tarragon complemented it perfectly. Serve this palate cleanser before the main course to give an elegant touch to a fancy dinner.

Brew a strong pot—2 cups (475 mL)—of high-quality black tea for a full 5-8 minutes. Pour into a shallow pan. Add 2 Tbsp. (30 mL) of honey and stir until dissolved. Chop 2 tsp. (10 mL) of tarragon lightly just before adding it to the tea.

Freeze for several hours, until the granita is semi-solid. Break it up with a spoon and stir. Return to the freezer and freeze until solid.

Place small bowls or stemmed glasses in the freezer for at least an hour before serving. To serve, scrape the granita with a spoon and place it in the frozen serving dishes. Serve immediately, or return to the freezer until ready to serve. Garnish with additional tea leaves. **−MC**

Tunisian-Scented Lamb
with Tomato Fig Butter

Serves 4

My sous-chef Ron Matusik and I enjoy a richly creative relationship. He took silver in a high-profile competition with this recipe. It truly is his creation, but I like to think it benefits from a sychronicity in our approach to food. I have used lamb sirloin here, which is a delicious and tender cut. If your butcher can't find it for you, you can substitute rack of lamb or lamb chops. Serve it with your favourite eggplant dish and perhaps some cous-cous, or try it with the White Bean & Garlic Mousse from *The Girls Who Dish* (page 137).

1/4 tsp.	ground allspice	1.2 mL
1/4 tsp.	freshly ground black pepper	1.2 mL
1 tsp.	ground cinnamon	5 mL
1 tsp.	salt	5 mL
3/4 tsp.	ground cumin	4 mL
4	5-oz. (140-g) pieces lamb sirloin	4
1 1/2 Tbsp.	olive oil	22.5 mL
1/4 cup	black Mission figs, packed	60 mL
1 Tbsp.	chopped fresh parsley	15 mL
1/4 cup	unsalted butter, softened	60 mL
1 cup	canned plum tomatoes, drained	240 mL
2 tsp.	unsalted butter	10 mL
1 Tbsp.	finely chopped shallots	15 mL
1 Tbsp.	unsalted butter	15 mL

Combine the allspice, pepper, cinnamon, salt and cumin. Place the lamb in a dish, rub it with the olive oil, then sprinkle the meat evenly on all sides with the spice mixture. Cover and refrigerate for several hours or overnight.

Soak the figs in hot water for 20 minutes. Drain and chop them finely. Fold the figs and parsley into the 1/4 cup (60 mL) softened butter.

Cut the tomatoes in half and remove the excess seeds. Chop the tomatoes finely. Heat a small saucepan over low heat and add the 2 tsp. (10 mL) butter. Sauté the shallots for 1 minute. Add the chopped tomatoes and cook for 5 or 6 minutes. Set aside.

Preheat the oven to 350° (175°C).

Heat a medium sauté pan over medium heat. Add the 1 Tbsp. (15 mL) unsalted butter. Sauté the lamb on all sides for 2 minutes per side. Remove from the heat and place the lamb on a small baking sheet. Roast for approximately 20 minutes, or until cooked to medium—an instant-read thermometer will register 150°F (65°C). Keep warm and allow to rest for 6-8 minutes before slicing.

Just before serving, reheat the tomato mixture and fold in the fig butter mixture. Season with salt and pepper and serve with the warm lamb.

tip: **Clarified Butter**

Melt butter in a heavy saucepan. Continue heating until the butter starts to crackle and sputter as the milk solids evaporate. Watch very carefully at this point; do not allow the butter to burn. When the sputtering subsides and the residue has just begun to toast, strain the butter immediately through cheesecloth or a very fine strainer, leaving the milk solids behind. –MC

MARGARET CHISHOLM

April's Pancetta Penne

Serves 4 to 6

Among the people who have influenced my cooking are the fabulous staff at Lesley Stowe Fine Foods. Our head pastry chef knows how to cook pasta the Italian way: al dente. This is my version of her most requested dish. Simple, delicious and comforting!

8 oz.	pancetta, cut in $1/2$ x $1/4$-inch (1.2 x .6-cm) strips	225 g
2 Tbsp.	olive oil	30 mL
8	small cloves garlic, slivered lengthwise	8
4	small shallots, slivered lengthwise	4
$1/2$ tsp.	finely chopped fresh rosemary	2.5 mL
$1/2$ tsp.	finely chopped fresh sage	2.5 mL
$1/4$ cup	extra virgin olive oil	60 mL
1 lb.	button mushrooms, sliced	455 g
4 cups	tomato sauce	950 mL
	large pinch chili flakes	
1 lb.	penne	455 g
	sea salt and freshly ground black pepper to taste	
$3/4$ cup	freshly grated Parmesan cheese	180 mL

Cook the pancetta for 2-3 minutes in a heavy saucepan over medium heat. Add the 2 Tbsp. (30 mL) olive oil, garlic and shallots; sauté until soft, 2-3 minutes. Add the rosemary and sage. Add the $1/4$ cup (60 mL) olive oil and the mushrooms. Turn up the heat to medium-high and stir until the mushrooms are golden. Add the tomato sauce and chili flakes. Reduce the heat and keep warm.

In a pasta pot, bring 4 quarts (4 L) of water with 1 Tbsp. (15 mL) of sea salt to a boil. Add the penne and cook until it's almost al dente, 2-3 minutes. Drain, reserving $1/2$ cup (120 mL) of the cooking water. Add the water to the sauce. Return the pasta and sauce to the pasta pot and cook over medium heat until the pasta is al dente, 2 minutes. Season with salt and pepper. Serve with Parmesan cheese.

Lemon Caper Pappardelle with Parmesan

If there is a golden rule that comes out of cooking in Italy, it's keep it simple and use the best ingredients of the season. This pasta dish epitomizes that philosophy. The pasta should always be the star of the show, with the other ingredients complementing it. This dish was inspired by time I spent in the kitchen with Marietta—Umberto Menghi's sister—at Villa Delia in Tuscany. Fresh herbs and produce brought in from the garden every morning, olive oil, wine from the property . . . how can one not be influenced?

Serves 6

1 lb.	pappardelle noodles (the best quality you can find)	455 g
1 Tbsp.	sea salt	15 mL
3/4 cup	extra virgin olive oil	180 mL
1/3 cup	capers	80 mL
1 Tbsp.	finely grated lemon zest	15 mL
2 Tbsp.	fresh lemon juice	30 mL
1/2 tsp.	sea salt	2.5 mL
1/4 tsp.	freshly ground black pepper	1.2 mL
1/2 cup	freshly grated Parmigiano-Reggiano	120 mL

Bring 4 quarts (4 L) of water and 1 Tbsp. (15 mL) sea salt to a boil in a large pot. Add the pasta and cook until al dente. Drain, letting water cling to the pasta. In a bowl, combine the olive oil, capers, zest, juice, salt and pepper. Toss the pasta back in the pot with the lemon caper mixture. Place in individual bowls and generously sprinkle the cheese on top of each serving.

Caren's Pork and Beans with Spicy Dressing

Serves 6

If pork and beans conjures up visions from the '50s, you are not alone. Sitting in front of a TV at dinner time, with cans providing most of your food prep, is not that appealing. We have come a long way, both in culinary tastes and dinner company! Try my version—it is nothing like the way it was.

1	1 1/2-inch (3.8-cm) piece fresh ginger, peeled and minced	1
2	medium cloves garlic, minced	2
2 Tbsp.	black bean sauce	30 mL
2 Tbsp.	soy sauce	30 mL
1 Tbsp.	rice vinegar	15 mL
1 Tbsp.	sesame oil	15 mL
1 tsp.	sugar	5 mL
1 tsp.	sea salt	5 mL
2 lbs.	pork tenderloin, trimmed of fat and membrane	900 g
1 lb.	thin green beans	455 g
1 recipe	Spicy Dressing	1 recipe
2 Tbsp.	black sesame seeds, toasted	30 mL
2 Tbsp.	white sesame seeds, toasted	30 mL

Combine the ginger, garlic, bean sauce, soy sauce, rice vinegar, sesame oil, sugar and salt and stir. Place the pork in a large reclosable freezer bag, pour the marinade in the bag and seal it. Place in the refrigerator and let it marinate for a least 4 hours or overnight.

Preheat the oven to 400°F (200°C). Remove the pork from the marinade and transfer it to a roasting pan. Roast for about 30–45 minutes, or until the internal temperature of the pork reads 165°F (75°C). Remove from the oven and set aside while you blanch the beans and make the dressing.

Blanch the beans in boiling water for about 1 minute, or until cooked but still firm. Transfer to a bowl of ice water to stop the cooking process. Drain and set aside.

To assemble, place the beans in a circular pattern around a medium-size platter. Slice the pork and lay it in an overlapping circular pattern in the centre of the beans. Drizzle the dressing evenly over the pork and beans. Garnish with the black and white sesame seeds.

Spicy Dressing

Makes $3/4$ cup (200 mL)

1 Tbsp.	grapeseed oil	15 mL
2 Tbsp.	sesame oil	30 mL
1 $1/2$ Tbsp.	dried chili flakes	22.5 mL
1	green onion, thinly sliced	1
$1/3$ cup	soy sauce	80 mL
2 Tbsp.	rice vinegar	30 mL
1 Tbsp.	sugar	15 mL
2 Tbsp.	sake	30 mL

Heat the oils in a small saucepan over medium-high heat. Add the chili flakes and green onion, and cook until soft and fragrant, about 3 minutes. Add the remaining ingredients and bring to a slow simmer. Cook for about 1 minute.

Now my inspirations
often come from a visit to
a local market.

I CONSIDER COOKING TO BE
PRIMARILY a craft. Artistry and
true originality are a rare thing.
As a chef I must rely on fine-
quality products, careful
preparation and thoughtful
combinations. It has got to
taste good. "Yum" is perhaps
the greatest compliment I
can receive.

Good food and learning
to cook was always of great
interest to me, and as a
teenager, my family were the
prime targets for my culinary
experiments. Even negative
inspiration can play a part;
who knows, perhaps my
brother's verdict of "terrible" for
my first attempt at spaghetti
was the spur that kept me
going on a career path that

I hadn't planned. (I always
thought I would go to univer-
sity and become a scientist,
but when I worked at a ski
lodge in the Rockies after high
school, they sent me into the
kitchen and somehow I never
got out.)

Some of the inspiring voices
I have learned from in my
career include Allan Smith, my
lead chef instructor, who said,
"Keep building upon the
flavours in a dish." Mara
Jernigan, a devoted chef I
met well into my career, said,
"Season everything; a tiny
pinch of salt and pepper
everywhere brings it all together." Andre Soltner, chef of
Lutece Restaurant in New York,
was quoted as saying, "It is

MARGARET CHISHOLM

true that you must learn to cook. It is also true that you must love to cook, have feeling, but that is not enough. Love without technique is no good. Technique without love is also no good."

Now my inspirations often come from a visit to a local market. When fresh hazelnuts from the Fraser Valley appear in the markets each fall, I make them the star at a dinner party for friends. I crack open a few, breathe in their nutty perfume, marvel at their buttery flavour and I attempt to imagine something new and original, or rework something tried and true.

Barbecued Spaghetti

Serves 4 to 6

Ever since my eldest brother described my first spaghetti recipe as "terrible," I have been trying to improve both myself and my spaghetti. This dish is perfect for hot summer days. Vary the vegetables or sausages, but keep these key points in mind: use lots of garlic and herbs, grill the vegetables until scorched, and serve it with plenty of slightly chilled cheap Chianti.

1 lb.	dried spaghetti or other pasta	455 g
2 Tbsp.	coarse salt	30 mL
2	red bell peppers	2
1/2 cup	extra virgin olive oil	120 mL
	sea salt and freshly ground black pepper to taste	
1	eggplant, sliced into 1/2-inch (1.2-cm) rounds	1
2	zucchini, sliced into 1/2-inch (1.2-cm) rounds	2
6–8	Roma tomatoes	6–8
8	cloves garlic	8
2	Italian sausages (optional)	2
1 cup	tomato-vegetable or tomato juice	240 mL
1/2 cup	chopped fresh basil	120 mL
1/4 cup	chopped fresh parsley	60 mL
1 cup	grated Parmesan cheese	240 mL

Preheat the barbecue to medium.

Bring a large pot of water to a boil over high heat. Add the pasta and coarse salt to the boiling water. Cook, stirring occasionally, until the pasta is tender, but still firm to the bite, approximately 9–11 minutes. Drain the pasta and return to the pot.

Cut the red peppers in half and remove the seeds and membrane. Place in a bowl and coat with 2 Tbsp. (30 mL) of the oil. Sprinkle with salt and pepper. Set the peppers aside. Toss the eggplant and zucchini with 2 Tbsp. (30 mL) of the oil and sprinkle with salt and pepper. Set aside.

Toss the tomatoes with 2 Tbsp. (30 mL) of the oil.

Place the unpeeled garlic cloves on the grill, off to the side. Place all the vegetables on the barbecue and grill until soft and scorched. If using sausages, add them to the grill as well. By now the garlic should be soft; if not, cook for a few more minutes.

Place all the vegetables on a large chopping board. Chop into 3/4-inch (2-cm) pieces. Cut the tips off the garlic cloves and squeeze the garlic out. Chop the sausages, if using, into 1/2-inch (1.2-cm) pieces.

Add the vegetables, sausages, garlic and juice to the pot of pasta and place over medium heat. Add the remaining 2 Tbsp. (30 mL) olive oil with the basil and parsley. Toss and heat through. Place in a heated serving dish, sprinkle with the cheese and toss.

Penne with Sautéed Artichokes, Tomatoes and Curry

Serves 4

This dish, made with fresh tomatoes and prawns, was one of my favourite pastas on the menu when I worked in a local Mediterranean restaurant. This version can be quickly prepared and does not require a trip to a specialty store or deli for hard-to-find ingredients.

1	14-oz. (398-mL) can artichokes, drained	1
1 Tbsp.	olive oil	15 mL
1	small onion, finely diced	1
2	cloves garlic, minced	2
2 tsp.	curry powder	10 mL
1	14-oz. (398-mL) can peeled plum tomatoes, with juice	1
1 cup	chicken stock or water	240 mL
1/2 lb.	dried penne	227 g
1/4 cup	fresh or frozen peas	60 mL
1/3 cup	whipping cream	80 mL
1/4 cup	chopped fresh basil leaves	60 mL
1/3 cup	grated Parmesan cheese	80 mL
	fine sea salt and freshly ground black pepper to taste	

Squeeze the artichokes by hand to remove any extra liquid. Cut them into quarters.

In a large saucepan, heat the olive oil over medium-high heat; when the oil is hot, add the artichokes. Cook, stirring occasionally, for 3–4 minutes, or until they begin to brown. Add the onion, garlic and curry powder. Cook, stirring often, for another minute. Add the tomatoes, breaking them into large pieces with the stirring spoon. Add the stock or water and bring to a boil; reduce the heat to medium-low and simmer for 10 minutes.

Bring a large pot of salted water to a boil. Add the penne to the boiling water and cook, stirring occasionally, until the pasta is tender but still firm to the bite, 9-11 minutes. Drain the pasta and return to the pot.

Add the peas and cream to the sauce and simmer for 1 minute. Stir in the fresh basil and Parmesan cheese. Season with salt and pepper. Toss the sauce with the pasta and serve in heated bowls.

tip: **Phyllo Pastry**

The best way to handle phyllo pastry is to work with it one sheet at a time. Keep the remainder of the pastry from drying out and becoming brittle by placing it under a dampened tea towel until needed. **–LS**

MARY MACKAY

Mushroom Ragoût
with Grilled Tuscan Bread

Serves 8

Vegetarians abound in the kitchen at Lesley Stowe Fine Foods; as a result, we are treated to all kinds of innovative vegetarian main courses. This mushroom ragoût garnished with sour cream and served with grilled Tuscan bread will satisfy even your most ardent beef eaters. Try serving it with penne, garlic mashed potatoes or gnocchi.

2	28-oz. (796-ml) cans whole Italian tomatoes, roughly chopped	2
1	large onion, diced	1
3	yellow bell peppers, cut in 3/4-inch (2-cm) cubes	3
6	large cloves garlic, slivered	6
1/4 cup	tomato paste	60 mL
1/2 tsp.	sea salt	2.5 mL
1/4 tsp.	freshly ground black pepper	1.2 mL
1/4 cup	Hungarian paprika	60 mL
1/3 cup	extra virgin olive oil	80 mL
8 cups	button, shiitake, oyster and portobello mushrooms, cut into 1/4-inch-thick (.6-cm) slices	2 L
1/2 cup	red wine	120 mL
1/2 cup	sour cream	120 mL
1 recipe	Grilled Tuscan Bread	1 recipe

Combine the tomatoes, onion, yellow peppers, garlic, tomato paste, salt, black pepper and paprika in a Dutch oven over medium-low heat. Simmer for 45 minutes, or until the onions are soft.

Heat the olive oil in a large sauté pan over medium-high heat. Add the mushrooms, season with salt and pepper and sauté until the mushrooms are soft and golden, approximately 10 minutes. Add the red wine and simmer for 2 minutes.

Add the mushrooms to the ragoût. Serve with a dollop of sour cream and a piece of Grilled Tuscan Bread.

Grilled Tuscan Bread

8	$^1/_2$-inch (1.2-cm) slices of dense artisan-style bread	8
$^1/_2$ cup	extra virgin olive oil	120 mL

Place the slices of bread on a hot grill or barbecue for 1–2 minutes per side—just enough to grill-mark them. Brush both sides lightly with olive oil and serve warm.

vegetables

(we love potatoes!)

vegetables

Grilled New Potatoes
with Smoky Poblano Vinaigrette

Serves 6

Potatoes are both a challenge and source of inspiration at Lesley Stowe Fine Foods, as we always have to have several versions on our constantly changing menu. One of the most interesting potato salads that has come out of our kitchens, this was originally inspired by our former executive chef, Liz Zmetana. I can't get enough of this smoky vinaigrette tossed with grilled potatoes! It makes a fabulous summer side dish with anything off the barbecue, particularly steak or chicken.

2 lbs.	small new potatoes	900 g
2 Tbsp.	extra virgin olive oil	30 mL
	sea salt and freshly ground black pepper to taste	
1	medium-sized red onion	1
1	red bell pepper	1
1	yellow bell pepper	1
1 recipe	Smoky Poblano Vinaigrette	1 recipe
1 Tbsp.	cilantro, coarsely chopped	15 mL

Preheat the oven to 375° (190°C).

Cut the potatoes in half and toss them in 1 Tbsp. (15 mL) of the olive oil. Season with salt and pepper. Place in a baking dish and roast for 15–20 minutes. Set aside.

Slice the onion into 1/2-inch (1.2-cm) rings, keeping each section intact. Cut the peppers in half, removing the membrane, seeds and stem.

LESLEY STOWE

Brush the onion and peppers with the remaining olive oil and sprinkle with salt and pepper. Place the vegetables on the grill, making sure the cut side of the potatoes and skin side of the peppers are facing the flames. Cook the onions very slowly over low heat. Grill on both sides so they cook through and get a chance to develop their sweetness. When the peppers are done, after about 15 minutes, remove them and cut them into triangles. When the onions are tender, transfer to a bowl with the peppers and potatoes. Toss with the vinaigrette and cilantro. Serve immediately.

Smoky Poblano Vinaigrette

Makes $^1/_2 - ^3/_4$ cup (120-180 mL)

3 Tbsp.	sherry vinegar	45 mL
$^1/_2$ tsp.	Dijon mustard	2.5 mL
$^1/_2$ tsp.	grainy mustard	2.5 mL
1 Tbsp.	chipotle purée (page 86)	15 mL
1	large clove garlic, chopped	1
$^1/_2$ tsp.	sea salt	2.5 mL
$^1/_4$ tsp.	freshly ground black pepper	1.2 mL
6 Tbsp.	extra virgin olive oil	90 mL

Whisk together the vinegar, mustards, chipotle purée, garlic, salt and pepper. Keep whisking and gradually pour in the olive oil. Taste and adjust seasoning, if desired.

Decadent Potatoes

Serves 6 to 8

My memory of great potatoes takes me back to New York City at the Union Square Café. My husband, Jose, ordered the garlic mashed potatoes with his main course and I reached over, as I always do, and stole a mouthful. They were unequivocally orgasmic. We ordered another serving, but in the meantime, I had a peek at their recently released cookbook. There was the recipe, in all its full cream glory. I hesitated for a moment but then dug in. It was well worth it. Here is my version of equally decadent potatoes. Just don't think about the cream—you are not eating the entire thing solo.

2 lbs.	red potatoes (about 6-8)	900 g
2 cups	heavy cream	475 mL
3 cups	grated Swiss Gruyère cheese	720 mL
	freshly ground Tellicherry pepper to taste	
	few gratings of nutmeg	
1/2 cup	grated Parmesan cheese	120 mL

Preheat the oven to 375°F (190°C). Lightly butter a 12-inch (30-cm) baking pan.

Scrub the potatoes well and blot them dry. Do not peel. Slice the potatoes very thinly on a mandoline. Divide into 3 equal portions. Pour about 1/4 cup (60 mL) of the cream into the bottom of the pan. Lay 1/3 of the sliced potatoes evenly in rows on top of the cream. Sprinkle 1/3 of the Gruyère on top of the potato slices. Pour 1/3 of the cream over this. Grind a little pepper and nutmeg over top. Repeat this process until all the potatoes, cream and Gruyère cheese is used up. You will have 3 layers.

Finish the dish by sprinkling the Parmesan cheese evenly over the top. Cover with foil and bake for 30 minutes. Remove the foil and bake an additional 20 minutes, or until the top is browned and the potatoes are tender.

Crème Fraîche Potato Gratin

It seems our love affair with potatoes is never-ending. This dish is a cousin to the classic Dauphinois. I urge you to to make it with the buttery yellow potatoes and crème fraiche just once, and see the difference this yummy potato makes and why crème fraiche is a world apart from sour cream.

Serves 6 to 8

3 lbs.	Yukon gold (or other yellow varieties) or russet potatoes	1.35 kg
2 cups	milk	475 mL
2 cups	water	475 mL
3	bay leaves	3
3	cloves garlic, minced	3
$1/2$ tsp.	salt	2.5 mL
	freshly ground black pepper to taste	
$1/8$ tsp.	freshly grated nutmeg	.5 mL
1 cup	crème fraîche (page 187) or whipping cream	240 mL
6 oz.	grated Parmesan or Grana Padano cheese	170 g

Preheat the oven to 375°F (190°C).

Peel the potatoes and slice them about $1/8$ inch (.3 cm) thick, using a chef's knife or mandoline. Place the potatoes in a large saucepan, add the milk, water, bay leaves, garlic and salt. Bring to a gentle boil over medium heat; stir gently a few times with a rubber spatula to avoid sticking. Simmer on low for about 10 minutes.

Butter a 2-quart (2-L) gratin or baking dish. Use a slotted spoon to transfer half the potatoes to the dish; season with pepper and nutmeg. Place half the cream and cheese over the potatoes. Repeat the layers, finishing with the cheese.

Bake until crisp and golden, about 1 hour.

GLENYS MORGAN

Crisp Yukon Gold Potatoes with Parsley and Garlic

Serves 2 to 4

When I was growing up, I never imagined I would choose cooking as a career. But after ten years of cooking in a variety of places, I finally decided to take it seriously and went to study in New York. My first day there, a version of this classic dish was on the menu. I knew I was on the right track.

2	large Yukon gold potatoes	2
3 Tbsp.	clarified butter (page 139)	45 mL
	sea salt and freshly ground black pepper to taste	
3 Tbsp.	fresh Italian parsley, chopped	45 mL
1	clove garlic, finely chopped	1

Bring a large pot of water to a boil. Meanwhile, peel the potatoes and cut them evenly into ^{1}/4-inch-thick (.6-cm) rounds. Add the potatoes to the boiling water and cook for 1 minute. Drain in a colander and refresh in cold water. Drain well and pat dry.

In a large sauté pan over medium-high heat, heat the clarified butter until hot but not smoking. Add just enough potatoes to fill the pan with one layer. Sauté the potatoes for 2-3 minutes, or until golden brown. Turn them over and brown the other side. Move the potatoes around in the pan to brown them evenly. Test the potatoes for doneness by poking them with a knife. They should be firm but tender.

Remove from the heat. Drain the potatoes on paper towel and transfer to a warm serving dish. Add the parsley and garlic, season with salt and pepper and toss well. Serve immediately.

MARGARET CHISHOLM

Porcini-Crusted Roasted Potatoes

The inspiration for this recipe came from none other than Margaret Chisholm, the person responsible for the fabulous Porcini-Dusted Sea Bass with Balsamic Brown Butter in *The Girls Who Dish– Seconds Anyone?* (page 124). Serve these potatoes with simple roasted meats. Dried porcini mushrooms can be crumbled and ground to a powder in a coffee grinder, but some specialty stores sell the powder.

Serves 4 to 6

3 Tbsp.	powdered porcini mushrooms	45 mL
3 Tbsp.	extra virgin olive oil	45 mL
$1/2$ tsp.	salt	2.5 mL
6	medium red potatoes, cut into 6 wedges each	6

Preheat the oven to 450°F (230°C).

In a large bowl, stir together the porcini, oil and salt. Add the potatoes and toss to coat them well on all sides. Spread the potatoes, one of the cut sides down, on a large, preferably non-stick baking sheet. Roast for 15 minutes. Turn the potatoes over and roast for 10–15 minutes longer, until fork-tender.

KAREN BARNABY

Potato Gratin with Wild Mushrooms

Serves 4 to 6

This dish reminds me of the pork chop and potato casserole my mother used to make with a can of mushroom soup. I prefer to sauté the pork chops separately and serve this gratin as a side dish. I use a variety of mushrooms to intensify the flavour of the potato gratin. If fresh mushrooms are not readily available, you could substitute soaked dried mushrooms.

1	clove garlic	1
1 1/2 Tbsp.	unsalted butter	22.5 mL
1 1/2 cups	sliced mixed mushrooms, such as chanterelle or portobello	360 mL
1 1/2 cups	half-and-half cream	360 mL
1 1/8 tsp.	fine sea salt	5.5 mL
1/8 tsp.	freshly ground black pepper	.5 mL
1 1/2 lbs.	Yukon gold potatoes, peeled and very thinly sliced	680 g

Preheat the oven to 350°F (175°C).

Cut the garlic clove in half and rub the bottom and sides of a shallow, 4-cup (950-mL) gratin dish with one half of the garlic clove. Mince the remaining garlic and set aside. Grease the dish with 1/2 Tbsp. (7.5 mL) of the butter.

In a large frying pan, heat the remaining 1 Tbsp. (15 mL) butter over medium-high heat; add the sliced mushrooms and cook, stirring often, for 4 minutes.

In a medium saucepan over medium-high heat, bring the cooked mushrooms, minced garlic, cream, salt, pepper and potatoes to a boil. Reduce the heat to medium and simmer for 3 minutes.

Pour the potato mixture into the gratin dish and level the top with a spatula. Bake the gratin until it's golden on top and the potatoes are soft, about 60–80 minutes. Allow the potatoes to cool for a few minutes before cutting and serving.

MARY MACKAY

Spring Asparagus
(page 174)

Coconut-Crusted
Prawns with Mango
Tango Sauce
(page 38)

Mahogany Glazed
Squash and Pears
(page 170)

Salmon à la Trudeau
(page 108)

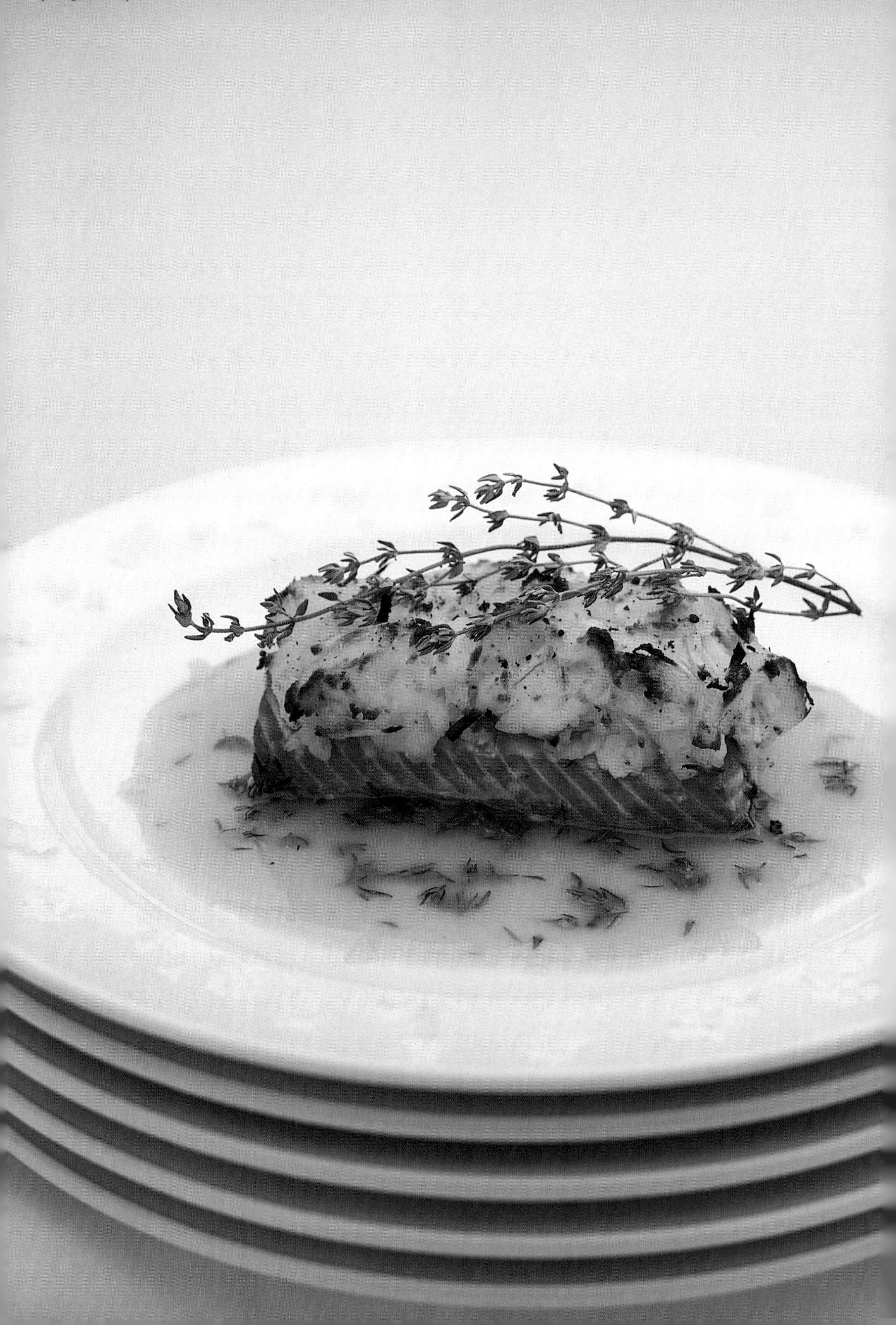

Wild Mushroom Ragoût

I would love to get my hands on the original newspaper clipping from the *Montreal Star* that was the inspiration for the recipe that follows. It was fall, probably around 1971. My dad and I used to clip recipes and collect them in a shoe box. I remember the moment I tasted this dish. It opened up a new world for me. What I would describe now as the layers of flavour, the balance and the richness wasn't part of my vocabulary then, I only knew I was really taken with the dish. Here it is, updated with wild mushrooms and fresh herbs. Serve this many ways: on sourdough toast that has been rubbed with garlic, in puff pastry cases, with pasta, polenta or gnocchi, or simply as an accompaniment to meat, fish or poultry.

Serves 4 as an accompaniment

2 Tbsp.	unsalted butter	30 mL
3/4 lb.	mixed mushrooms, button or wild, cleaned and sliced	340 g
1/3 cup	white wine	80 mL
2 Tbsp.	chopped shallots	30 mL
2	tomatoes, peeled, seeded and diced	2
1 1/2 tsp.	chopped fresh thyme	7.5 mL
3/4 tsp.	sea salt	4 mL
1/2 tsp.	freshly ground black pepper	2.5 mL

Warm a medium skillet over medium heat. Add the butter and when the foam subsides add the mushrooms. Cook and stir until the juices are released from the mushrooms. Continue to cook until the juices are reduced to a glaze. The mushrooms should just yield to light pressure, but retain their shape. This will take 3–4 minutes. Add the white wine, shallots, tomatoes and thyme. Cook for 3–4 minutes, or until the liquid is reduced and the mixture is thick. Season with salt and pepper.

MARGARET CHISHOLM

Roast Portobello Mushrooms
over Polenta Cakes

Serves 4 to 6

Meaty portobello mushrooms have become readily available. They are a great accompaniment to red meats, and their mild flavour and dense texture make them an excellent vegetarian dish. They can also be diced and served at room temperature as a salad garnish or a topping for bruschetta. Polenta cakes can be prepared a day ahead and then finished in the oven just before serving. Use them as salad garnishes or as a starch with almost any meat, poultry or fish. Cut quite small and topped with roast tomato confit or black olive tapenade, they make a great appetizer.

3 Tbsp.	olive oil	45 mL
1	clove garlic, minced	1
2 Tbsp.	balsamic vinegar	30 mL
1 tsp.	fresh thyme leaves	5 mL
6	large portobello mushrooms	6
1 tsp.	salt	5 mL
1 tsp.	freshly ground black pepper	5 mL
1 recipe	Polenta Cakes	1 recipe

Preheat the broiler to high.

Whisk the oil, garlic, vinegar and thyme in a medium bowl. Remove and discard the stems from the mushrooms. Toss the mushrooms with the oil and vinegar mixture until they are well coated. Remove the mushrooms from the marinade and season with salt and pepper. Place them on a baking sheet, gill side up and roast under the broiler for 2 minutes. Turn the mushrooms over and roast for 1 more minute. Turn off the broiler and set the oven temperature to 350°F (175°C). Move the mushrooms to a lower shelf and continue to bake them for 5 more minutes.

DEB CONNORS

You can also do these on the barbecue. Cook them over medium-high heat for 1 minute per side, then reduce the heat to medium and grill for 3–4 minutes.

Serve roasted mushrooms over top of the Polenta Cakes.

Polenta Cakes

Serves 6 to 8

2 cups	chicken stock	475 mL
3/4 cup	whipping cream	180 mL
1	clove garlic, minced	1
3/4 cup	finely ground yellow cornmeal	180 mL
2 Tbsp.	butter	30 mL
2 Tbsp.	grated Parmesan cheese	30 mL
	salt and freshly ground black pepper to taste	

Bring the chicken stock, cream and garlic to a boil in a medium saucepan over medium-high heat. Whisking constantly, slowly pour in the cornmeal. Return the pan to a simmer and continue cooking the polenta over very low heat, stirring occasionally, for 6–8 minutes. When the polenta is thick and smooth, remove the pan from the heat and stir in the butter and Parmesan. Season with salt and pepper.

Line a baking pan with plastic wrap or parchment paper. While the polenta is still warm, pour it into the pan and spread it in an even layer 3/4 inch (2 cm) thick. Let cool, cover, placing plastic wrap directly onto the surface of the polenta so it doesn't dry out, and refrigerate for several hours or overnight.

When the polenta is cold, turn it out onto a cutting board. Remove the plastic wrap or parchment. Using a sharp knife or a cutter, cut the polenta into the desired shapes. To reheat, place the polenta on a baking sheet in a preheated 500°F (260°) oven for 6 minutes or gently sauté in a small amount of olive oil over medium heat for 3 minutes per side.

Garlicky Yam Puddings
with Gorgonzola Cream

Serves 6

These yam puddings are a perfect make-ahead solution for the holiday table. Unmould them on individual plates or a platter; drizzle them with maple syrup or sprinkle them with thyme. They can be made up to 2 days ahead and rewarmed in a low oven.

2 lbs.	yams	900 g
1	bulb roasted garlic, cloves removed and mashed	1
1/8 tsp.	freshly grated nutmeg	.5 mL
3 Tbsp.	dark brown sugar or maple syrup	45 mL
1/2 cup	whipping cream	120 mL
1/2 tsp.	salt	2.5 mL
	freshly ground black pepper to taste	
2	large eggs, beaten	2
2 Tbsp.	butter	30 mL
3/4 cup	whipping cream	180 mL
1/4 cup	Gorgonzola dolce, or any mild blue cheese	60 mL
2 Tbsp.	minced chives	30 mL
2 Tbsp.	grated Parmesan cheese	30 mL
	fresh chive sprigs	
	shaved Parmesan cheese (optional)	

Preheat the oven to 400°F (200°C). Place the whole yams on a parchment or foil-lined baking sheet. (Lining saves the pan from the sugary juices that burn during baking.) Bake the yams until fork-tender and caramelized brown spots appear on the skin. Cool, peel and mash the yams. You will need 1 1/2 cups (360 mL) of mashed yam. This step may be done ahead. To continue, lower the oven temperature to 350°F (175°C).

GLENYS MORGAN

Transfer the yams to the food processor bowl. Add the roasted garlic, nutmeg, sugar or maple syrup, the $1/2$ cup (120 mL) whipping cream, salt and pepper. Purée until very smooth. Taste the mixture and adjust the seasonings before adding the eggs. The flavour should be a nice balance of garlic and yam but not too sweet. Add the eggs and process until smooth.

Butter or spray the inside of six $1/2$-cup (120-mL) ramekins. Spoon the yam mixture into the ramekins, filling them $3/4$ full. (The mixture will expand during cooking.) Tap the ramekins on the counter to settle the purée and eliminate air bubbles. Set them in a larger baking pan and add enough boiling water to come $2/3$ of the way up the sides of the ramekins. Bake for 20–45 minutes: the time will vary depending on the size of ramekins and their spacing in the water bath. The puddings are done when they are very firm to the touch and pull away slightly from the side of the ramekin.

Remove from the oven and let stand for 10 minutes before unmoulding. The hot water bath may be used to keep them warm until serving time.

While the puddings are baking, make the sauce. Melt the butter and whisk in the $3/4$ cup (180 mL) cream. Cook until the butter is incorporated into the cream; continue to reduce until the sauce coats the back of a spoon. Break the cheese into small pieces and whisk gently into the cream, leaving small pieces whole in the sauce for extra flavour. Add the minced chives and Parmesan to the sauce just before serving. Adjust the seasonings. For a more subtle cheese flavour, the Parmesan can be omitted.

The rested puddings should unmould easily. Leaving them inverted on the plate for a minute helps. For individual servings, unmould each ramekin in the centre of the plate. Drizzle the sauce around it and garnish with chive sprigs and a shaving of Parmesan, if desired.

Honey-Glazed Carrots
with Shallots and Ginger

Serves 4

These carrots are a welcome addition to any meal: simple and delicious they will complement beef and lamb.

1 Tbsp.	olive oil	15 mL
2 Tbsp.	butter	30 mL
3	shallots, diced	3
1 Tbsp.	peeled minced fresh ginger	15 mL
1 Tbsp.	honey	15 mL
1 lb.	baby carrots, scrubbed	455 g
	salt and freshly ground black pepper to taste	

Preheat the oven to 400°F (200°C).

Heat the oil in a large sauté pan over medium-high heat. Add the butter, shallots, ginger and honey and cook for 3–4 minutes. Add the carrots and cook for 2–3 minutes, tossing to make sure the carrots are well glazed. Season with salt and pepper to taste. Transfer the carrots to a shallow baking dish in an even layer; do not overcrowd them. (You can prepare these carrots ahead to this point; just make sure they are at room temperature before you finish them in the oven.)

Place in the oven and roast, stirring occasionally, until they are lightly browned and tender, approximately 10–15 minutes. The cooking time will vary depending on the size and thickness of the carrots, so be sure to check them partway through the cooking process.

DEB CONNORS

Carrots and Parsnips Roasted
with Indian Spices

Roasted vegetables have always been comfort food to me, and with the influence of several East Indian women in my kitchen, I've opened my mind to experimenting with their spices in non-traditional ways. These vegetables are superb when served with roast chicken or pork.

Serves 6

2 lbs.	baby carrots, washed and cut into 1-inch (2.5-cm) lengths	900 g
2 lbs.	parsnips, peeled and cut into 1-inch (2.5-cm) lengths	900 g
2 Tbsp.	olive oil	30 mL
1 1/2 tsp.	ground ginger	7.5 mL
1/2 tsp.	freshly ground black pepper	2.5 mL
1/2 tsp.	ground allspice	2.5 mL
1/2 tsp.	ground nutmeg	2.5 mL
1/2 tsp.	ground mace	2.5 mL
1/2 tsp.	ground cardamom	2.5 mL
1/4 tsp.	ground cinnamon	1.2 mL
1/4 tsp.	ground turmeric	1.2 mL
1/4 tsp.	ground coriander	1.2 mL
	pinch sea salt	

Preheat the oven to 375°F (190°C). Place the carrots and parsnips in a bowl. Combine the remaining ingredients, pour over the vegetables and toss to coat. Place in a single layer in a baking pan and roast for 20–35 minutes, until the vegetables are golden and tender.

And my original inspiration, my mom, is also part of the new generation of enthusiasts.

EVERY DAY THAT I TEACH, I meet people who love to cook, some whose goal is to own a restaurant and others who cook at home but want the skills of a chef. The media often paints a gloomy picture of how people are eating today, but I'm inspired by the new generation of cooks— ages sixteen to sixty—who have made our cooking classes full, exciting and rewarding. Complete strangers crowd around hot stoves and within a few hours are passionately describing—bite by bite—great meals or favourite recipes. We talk about food and cooking and eating and realize inspiration comes in many forms.

In Saskatchewan, seasons are not subtle, but they define life on the farm. The family farm, bustling markets in far-away places, ethnic dishes cooked with passion, cemented my view there shouldn't be one long supermarket season.

Jacques and Julia were always there for technique, but I also have my gurus of seasonal cooking. Alice Waters is an obvious choice. From her legendary Chez Panisse in Berkeley, she's inspired people to search out what's best, what's local, and what's right for the season. While Alice Waters was quietly leading the revolution, on the other side of the continent was author Perla Meyers. Her books, *Seasonal Kitchen* and *The Peasant Kitchen*, were the first ones that really spoke to me. Classically trained, speaking seven languages, she moved through the kitchens of tiny little gems in the countryside of Spain and Italy, and then into the kitchens of Bocuse and Troigros. Her style was really a philosophy: cook and eat in tune with the seasons. Eventually I met Perla; many books, classes and years later, we've cooked together often.

GLENYS MORGAN

GLENYS MORGAN

These two very different women in food inspired me or perhaps just reinforced a lesson in nature from the farm.

Tuned to the season, I leave the East Vancouver farmer's market each summer Saturday with amazing lettuces, many varieties of onions, and tomatoes in all the colours of the rainbow. I think of Alice Waters. As the summer fades and the growing season is coming to an end, I think of Perla who is looking forward to the kale, squash and apples. But it's the farmer's daughter in me that hopes all the vendors, with their crops for sale, can still grow all this amazing food and stay on the farm.

And my original inspiration, my mom, is also part of the new generation of enthusiasts. Not sitting on her laurels—or raspberry canes—she constantly tries new dishes and techniques, reads food columns, clips and cooks. She has her grandchildren in the kitchen, so I think the world of food is still a bright warm place when it starts around the stove at home.

Mahogany Glazed Squash and Pears

Serves 6 to 8

This dish began in the kitchens of Tra Vigne in Napa Valley, where they make it with an Italian favourite—sugar pumpkin. I loved the look of the glazed vegetables just before they were mashed, like big glossy gems, and so at home in my kitchen the recipe ends on a different note. Squash and pears are perfect companions, especially with a sweet glaze, or each could fly solo. Serve with grilled pork or a veal chop, or update the offerings when feasting on ham or turkey.

3 lbs.	butternut squash	1.35 kg
1 lb.	underripe pears	455 g
1/2 tsp.	salt	2.5 mL
	freshly ground black pepper to taste	
1/2 cup	unsalted butter	120 mL
2 Tbsp.	sugar	30 mL
1/4 cup	balsamic vinegar	60 mL
1/4 cup	molasses	60 mL

Preheat the oven to 400°F (200°C).

Peel and seed the squash. Peel and core the pears. Cube both into generous 1-inch (2.5-cm) pieces. (Butternut squash and pears are easily peeled with a T-shaped peeler with the blade mounted across the top. Peel in long smooth strokes from top to bottom. To peel squash easily and safely, halve it and place the cut side down on the cutting board before peeling. Mix the squash and pears in a large bowl. Season with salt and pepper.

Choose a baking dish large enough to hold the squash and pears in a single layer. (Avoid metal; the vegetables stick.) Spray or grease the baking dish.

Make a "brown butter" by heating the butter in a small saucepan over high heat until bubbly and lightly browned. Add the remaining ingredients, cooking for 1 minute to form a syrup.

Pour the glaze over the squash and pears, tossing to coat. Turn them into the baking dish. (They can be prepared to this point a day ahead and held in the fridge until baking time.) Place in the preheated oven. Bake until tender, about 30–40 minutes, turning occasionally with a rubber spatula. Serve very hot or just warm. Any leftovers may be mashed for another meal.

---

tip: **The Daily Grind**

Starbucks has changed the way we think about the daily grind, and we should also think about what we grind in our pepper mills. We choose our coffee beans by variety for flavour, so make the leap and buy peppercorns by pedigree. Put Tellicherry black, Montuk white or Sawasdee in your grinder. Remember, the layering of great ingredients in a dish is what builds great flavour. Details count! –GM

GLENYS MORGAN

Brussels Sprouts with Cream Cheese, Toasted Almonds and Nutmeg

Serves 4 to 6

In the Mackay household, Thanksgiving and Christmas dinners have to include steamed Brussels sprouts: they are my Dad's favourite. But it was apparent from the lack of little green cabbages on everyone's plate at the dinner table that none of the other family members shared Dad's appreciation for Brussels sprouts. That changed after my sister, Tede, introduced this recipe to the family. The cream cheese cuts the bitterness of the sprouts, and nutmeg is a perfect spice for the holidays. But Dad still prefers his simply steamed and tossed with butter!

20	medium Brussels sprouts, washed and cut in half	20
1/4 cup	cream cheese, cut into 8 pieces	60 mL
	fine sea salt and freshly ground black pepper to taste	
1/4 cup	slivered almonds, toasted	60 mL
	freshly grated nutmeg to taste	

Steam the Brussels sprouts in a medium pot until crisp-tender, about 5 minutes. Drain and toss immediately in the cream cheese. Season with salt and pepper. Transfer the sprouts to a warm serving dish and sprinkle with the almonds and nutmeg.

Broiled Spinach with Four Cheeses

Serves 4

My mother used to make a creamy and cheesy salmon casserole that used a lot of frozen spinach. It was the inspiration for Salmon Bake with Sour Cream, Bacon & New Red Potatoes in *The Girls Who Dish* (page 76). That creamy, cheesy spinach was also the inspiration for this quickly assembled dish with an updated Italian touch.

1/2 cup	grated Asiago cheese	120 mL
1/2 cup	ricotta cheese	120 mL
1/4 cup	crumbled Gorgonzola or blue cheese	60 mL
2 Tbsp.	freshly grated Parmesan cheese	30 mL
2 Tbsp.	chopped fresh basil	30 mL
1	large egg yolk	1
2 Tbsp.	extra virgin olive oil	30 mL
2	cloves garlic, minced	2
2	10-oz. (285-g) packages fresh spinach, coarsely chopped	2

Preheat the broiler to medium. Lightly butter an 11 x 7 x 2-inch (28 x 18 x 5-cm) baking dish.

Mix the cheeses, basil and egg yolk in a large bowl. Heat the oil in a large pot over medium-high heat, then add the garlic. Stir and cook for a minute until fragrant. Pour into the prepared baking dish. Add the spinach to the pot and sauté until wilted. Transfer the spinach to a strainer; drain well. Place the spinach over the garlic in the baking dish. Toss to coat it with the oil and spread it out evenly. Sprinkle with the cheese mixture. Broil until the cheese is golden on top and the spinach is heated through.

KAREN BARNABY

Spring Asparagus

Serves 6

Nothing denotes the arrival of spring better than tender shoots of young asparagus. Simplicity in preparation brings out the best in most tender veggies, and this is especially true of asparagus. Serve this dish hot or cold, depending on your menu.

1 lb.	tender young asparagus	455 g
2 Tbsp.	4-year-old balsamic vinegar	30 mL
1/3 cup	extra virgin olive oil	80 mL
	fleur de sel to taste	
	freshly ground Tellicherry pepper to taste	
1/2 cup	Parmesan cheese shavings	120 mL

Bring a pan of salted water to a boil. Snap off the woody ends of the asparagus. Place the spears in the boiling water for 2-3 minutes, or until they are still crisp. Do not overcook. Immerse immediately in ice water to stop the cooking process. Wrap the asparagus in a kitchen towel and chill until serving time.

To serve, place the asparagus on a serving plate and drizzle with the balsamic vinegar and oil. Sprinkle with the fleur de sel and pepper. Garnish with Parmesan shavings.

Cherry Tomatoes in Brown Butter Sauce

Cherry tomatoes used to scare me. I ate a rotten one as a child and found the way it popped in my mouth utterly vile. Now, I can barely wait for cherry tomato season. There are so many wonderfully sweet varieties available now, especially from organic growers, that I just eat them plain. This is a quick dish that I make with my new-found love. Using different varieties of cherry tomatoes is especially attractive.

Serves 4

3 Tbsp.	unsalted butter	45 mL
3 cups	cherry tomatoes, stems removed	720 mL
$1/4$ cup	chopped fresh parsley	60 mL
$1/4$ cup	fresh basil leaves	60 mL
1 Tbsp.	lemon juice	15 mL
	sea salt and freshly ground black pepper to taste	

Melt the butter over medium heat, stirring constantly until brown. Add the tomatoes and cook about 3 minutes, or until the skins pop, stirring constantly. Add the parsley, basil and lemon juice. Season with the salt and pepper and serve immediately.

KAREN BARNABY

desserts

Chocolate Macadamia Nut Cookies

Makes about 24 cookies

I could never live in Hawaii; I have no control when it comes to chocolate-covered macadamia nuts. These cookies are a real treat; the texture is somewhere between a soft cookie and a brownie. The fresh lime zest gives the cookies a tropical flare. It is important to use good-quality chocolate as it is the main flavour base of the cookie.

6 Tbsp.	unsalted butter, at room temperature	90 mL
1 tsp.	lime juice	5 mL
1/2 lb.	good-quality bittersweet chocolate, finely chopped	227 g
2	large eggs	2
1/3 cup	granulated sugar	80 mL
1/4 cup	brown sugar, packed	60 mL
2/3 cup	unbleached all-purpose flour	160 mL
1/4 tsp.	baking soda	1.2 mL
1/4 tsp.	fine sea salt	1.2 mL
1 1/2 tsp.	lime zest	7.5 mL
1/2 cup	chopped toasted macadamia nuts	120 mL

Preheat the oven to 350°F (175°C). Line 2 17 x 11-inch (43 x 28-cm) baking sheets with parchment paper. Then place each sheet inside a second baking sheet to insulate. Altogether, you will use 4 baking sheets.

In a medium saucepan, melt the butter and lime juice over medium heat. Reduce the heat to low and add the chocolate, stirring until completely melted.

In a large bowl, beat the eggs and both sugars until thickened, about 5 minutes. Fold in the chocolate mixture. In a separate bowl, stir together the flour, baking soda, salt and lime zest. Mix the flour into the batter along with the macadamia nuts. Refrigerate the batter for 10-15 minutes, until firm enough to scoop.

MARY MACKAY

Scoop heaping tablespoons (15 mL) of the batter onto the prepared pans, placing them about 1 1/2-2 inches (3.8-5 cm) apart. You will get 12 cookies per baking sheet. Bake until set on the outside but still soft on the inside, about 10-12 minutes. Do not overbake. Slide the baking paper off the baking sheets and onto racks to cool the cookies.

---

quick bite: **John's Marshmallow Liqueur Cups**

My husband, John, and I had the opportunity to work together on food and beverage recipes using a Caribbean liqueur called Ponche Kuba. John prepared recipes for martinis and shooters with the liqueur, while I made milkshakes and cheesecakes with it. Our favourite recipe was John's marshmallow liqueur cups. We place these toasted marshmallow cups on top of bowls of ice cream and fill them with liqueur.

Thread a large marshmallow on the end of a metal skewer, piercing the marshmallow 3/4 of the way down. Toast the marshmallow over the stovetop burner, gently turning so that all sides turn golden brown. Do not drop the marshmallow on your stovetop or you will be in for a big cleanup. Let the marshmallow cool long enough so that it can be held comfortably in your hand. Hold the marshmallow in one hand and gently pull the skewer out of the marshmallow with your other hand, being careful not to squeeze the marshmallow. Some of the soft centre and the top of the marshmallow should pull out with the skewer, leaving you with a toasted marshmallow cup. The cups can be prepared ahead of time and stored for up to 2 weeks in a plastic container when completely cool. –MM

Chocolate, Cinnamon and Pumpkin Seed Cookie Brittle

Makes about 24 pieces

A new-found fondness for the combination of cinnamon and chocolate, plus a lifelong affair with my mother's chocolate chip cookie brittle, led to this recipe. I like to use pieces of cinnamon ground coarsely in a mortar and pestle for little explosions of sweet cinnamon flavour. My pet name for this is Mayan cookie brittle.

1/4 lb.	unsalted butter	113 g
1/2 tsp.	coarsely ground cinnamon	2.5 mL
1 oz.	unsweetened chocolate, chopped into small pieces	28 g
1/2 cup	sugar	120 mL
1/4 tsp.	vanilla extract	1.2 mL
1	egg	1
1/3 cup	flour	80 mL
1/2 cup	hulled pumpkin seeds	120 mL

Preheat the oven to 375°F (190°C).

Melt the butter in a large saucepan over medium heat. Add the cinnamon. Remove from the heat, add the chocolate and stir until melted. Stir in the sugar, vanilla, and egg until smooth. Stir in the flour and beat well. Immediately scrape the batter into a 10 x 15 x 1-inch (25 x 38 x 2.5-cm) jelly-roll pan lined with parchment paper. Spread the batter into a thin, even layer across the pan. Scatter the pumpkin seeds evenly over the top.

Bake for 10 minutes, or until the cookie is just set. Remove from the oven and cool on a rack for 20 minutes. Cut or break into 2 dozen rough-shaped pieces, as you would a nut brittle.

Maple Hazelnut Shortbread

Hazelnuts, or filberts as they are sometimes called, are grown in the Fraser Valley near Vancouver. Every year I look forward to buying a supply of fresh hazelnuts from the "hazelnut man" who sets up a stall at the market for several weekends in the fall. As with most nuts, they are best bought from a specialty shop or health food store when out of season. Maple sugar may be found in health food or specialty shops. If you can't find it, this recipe is still splendid when made with icing sugar. If you are using icing sugar, add 1 tsp. (5 mL) vanilla extract to the butter.

Makes 20 cookies

1 cup	all-purpose flour	240 mL
3 Tbsp.	cornstarch	45 mL
1/2 tsp.	salt	2.5 mL
3/4 cup	unsalted butter	180 mL
1/4 cup	maple sugar or icing sugar	60 mL
1/2 cup	chopped hazelnuts or filberts	120 mL

Combine the flour, cornstarch and salt. Set aside. Beat the butter by hand or in a mixer until light and soft. Beat in the sugar until the mixture is light and fluffy.

Mix in the flour mixture 1/3 at a time. Mix in the chopped nuts. Turn the dough out on a lightly floured work surface and knead very lightly for 15 seconds, or just until the dough comes together and is smooth. Chill for 30 minutes.

Preheat the oven to 325°F (165°C).

Turn out on a lightly floured work surface. Roll out 3/8-inch (1 cm) thick. Cut into 3-inch (7.5-cm) triangles. Place on a cookie sheet. Bake until firm and pale tan in colour, about 8–10 minutes. Cool on a rack. These cookies will keep for 2–3 weeks in an airtight container.

MARGARET CHISHOLM

Triple C (Chocolate Chocolate Cherry) Cookies

Makes 2 dozen cookies

Who doesn't love cookies and chocolate? This is a double whammy! A luxurious version of the old classic we all grew up with, it will satisfy the chocolate cravings of the grown-up children in your house!

1 1/4 cups	flour	300 mL
2 Tbsp.	unsweetened Dutch cocoa powder	30 mL
1 tsp.	baking powder	5 mL
1/2 tsp.	salt	2.5 mL
1/2 cup	unsalted butter	120 mL
12 oz.	high-quality bittersweet chocolate, coarsely chopped	340 g
1/2 cup	sugar	120 mL
1 tsp.	vanilla extract	5 mL
3	eggs	3
1/2 cup	toasted pecan halves	120 mL
3/4 cup	dried cherries	180 mL
8 oz.	high-quality bittersweet chocolate, coarsely chopped	227 g

Preheat the oven to 350°F (175°C). Butter an 11 x 17-inch (28 x 43-cm) baking sheet or line it with parchment.

Sift the flour, cocoa, baking powder and salt into a large bowl. In the top of a double boiler melt the butter and the 12 oz. (340 g) chocolate. Stir until melted and smooth. Remove the chocolate from the heat and stir in the sugar and vanilla. Add the eggs one at a time. Stir in the flour mixture, pecans, dried cherries and the remaining 8 oz. (227 g) of chocolate chunks. Chill the dough for a minimum of 30 minutes.

Drop tablespoons (15 mL) of dough 1 inch (2.5 cm) apart onto the prepared baking sheet. Bake for 6–8 minutes, until just set.

Clara's Date and Oatmeal Cookies

These cookies hail from my Mom's kitchen. Monday was her baking day, and these were one of my favourites. The best part about these cookies is that they are never soggy because you sandwich together only what you will eat. They hold up for 6 hours before they soften. The filling keeps for a week in the fridge.

Makes 2 dozen filled cookies

1 cup	unsalted butter	240 mL
1 cup	brown sugar, packed	240 mL
2 cups	unbleached all-purpose flour	475 mL
1 1/2 cups	quick-cooking oats	360 mL
1 tsp.	baking soda	5 mL
	pinch sea salt	
3 Tbsp.	cold water	45 mL
3 1/2 cups	pitted dates, chopped	840 mL
1 1/2 cups	hot water	360 mL
1 Tbsp.	fresh lemon juice	15 mL

Preheat the oven to 350°F (175°C). In a large bowl, beat the butter and sugar together. In a separate bowl, stir together the flour, oats, soda and salt. Add the flour mixture to the butter mixture, beating until combined. Gradually beat in the water. Using your hands, gather the dough together, working it into a ball.

Cut the dough into quarters. On a lightly floured surface, roll each portion of dough to 1/4 inch (.6 cm) thick. Cut the dough into rounds with a 2 1/4-inch (5.6-cm) cookie cutter. Place on a parchment-lined cookie sheet and bake for about 12–15 minutes, or until golden.

In a saucepan, combine the dates, water and lemon juice. Bring to a boil, reduce the heat to medium-low and cook for 10 minutes, or until the dates are spreadable. Transfer to a bowl to cool. To serve, spread one side of the cookie with 4 tsp. (20 mL) of filling. Place another cookie on top and press gently. If you like them crispy, eat them right away; if you prefer a softer cookie, let them stand for a few hours.

CAREN MCSHERRY-VALAGAO

Rhubarb Coconut Cake

Makes one
9-inch (23-cm)
cake

This easy to prepare and extremely delicious cake includes two of my favourite flavours. The recipe comes from my friend Mara Jernigan on Vancouver Island, who says: "When I see a cake recipe that calls for buttermilk I know it will be moist and have a slightly tangy, rich taste. This one from my late mother combines spring rhubarb with a coconut topping for a most unusual but delicious match. Serve with crème fraiche or ice cream and a glass of Italian Muscato or a B.C. dessert wine."

2 cups	rhubarb chopped into $^1/_2$-inch (1.2-cm) pieces	475 mL
1 $^1/_2$ cups	granulated sugar	360 mL
$^1/_2$ cup	brown sugar	120 mL
$^1/_4$ cup	butter	60 mL
$^1/_2$ cup	shredded, sweetened coconut	120 mL
1 tsp.	ground cinnamon	5 mL
2 cups	flour	475 mL
1 tsp.	baking soda	5 mL
$^1/_2$ cup	butter	120 mL
1	egg	1
1 cup	buttermilk	240 mL

Preheat the oven to 350°F (175°C). Butter a 9-inch (23-cm) springform or square cake pan.

Sprinkle the rhubarb with 2 Tbsp. (30 mL) of the granulated sugar and set aside while you prepare the rest of the cake. In a separate bowl, mix the brown sugar, $^1/_4$ cup (60 mL) butter, coconut and cinnamon by hand until chunky but well incorporated. Set aside.

Whisk together the flour and baking soda. Combine the remaining white sugar and the $^1/_2$ cup (120 mL) butter. Beat with an electric mixer on medium-high speed for 3-5 minutes, or beat by hand until light and fluffy. Add the egg and buttermilk and mix on slow speed, scraping down the sides of the bowl. Add the flour mixture and mix on slow speed just until incorporated. Fold in the rhubarb along with any juice in the bowl, mixing it in with a rubber spatula. Pour the batter into the prepared pan and sprinkle with the brown sugar-coconut topping.

Bake for 30-40 minutes, or until a toothpick inserted in the centre comes out clean. Cool on a rack before cutting.

tip: **Candied Orange Zest**

These make a pretty garnish for desserts. Using a vegetable peeler, remove the outer skin of 2 oranges, keeping the pieces as long and large as possible. Using a very sharp knife, cut the orange peelings into a very fine julienne. In a small saucepan combine 1 cup (240 mL) sugar and 1 cup (240 mL) water. Bring it to a boil, reduce the heat to medium-low, add the julienned zest and simmer for 10 minutes. Strain the zest, using a fine sieve. (You can cool the sugar syrup and store it in the refrigerator to use again; it will keep for 3-4 weeks.) In a medium bowl toss the orange zest with $^1/_4$ cup (60 mL) granulated sugar. Let it sit in the sugar, tossing several times, for 10 minutes. Shake the zest in a sieve to remove the excess sugar. Spread the zest on a plate and let dry 2-3 hours. When it has dried you can store it in a tightly covered container at room temperature for up to 1 week. **–DC**

KAREN BARNABY

Nana's Hot Milk Cake

**Makes one
8-inch (20-cm)
square cake**

This was a favourite dessert made by my grandmother. It is a simple white sponge cake topped with baked coconut icing. The cake was named for the scalded milk that was added to the batter. My father always felt that it was not a true hot milk cake unless you literally poured hot milk over the finished cake, so Nana always served us big wedges in a bowl with warm milk poured over the top. You could substitute Deb's Orange Crème Anglaise in *The Girls Who Dish* (page 173), served warm, for the hot milk.

1/2 cup	whole milk	120 mL
3 Tbsp.	unsalted butter	45 mL
2	large eggs	2
1 cup	granulated sugar	240 mL
1/4 tsp.	lemon zest	1.2 mL
1 tsp.	vanilla extract	5 mL
1 cup	all-purpose flour	240 mL
1 tsp.	baking powder	5 mL
1/8 tsp.	fine sea salt	.5 mL
2 1/2 Tbsp.	whipping cream	37.5 mL
3 Tbsp.	unsalted butter	45 mL
5 Tbsp.	brown sugar	75 mL
1/2 cup	unsweetened shredded coconut	120 mL

Preheat the oven to 325°F (165°C). Butter and flour an 8-inch (20-cm) square baking pan.

In a small saucepan over medium-high heat, scald the milk and the 3 Tbsp. (45 mL) butter. Set aside to cool.

Place the eggs and granulated sugar in the bowl of a heavy-duty mixer (or use a hand mixer). Using the whisk attachment, beat on medium speed until pale and thickened, about 4 minutes. Whisk in the lemon zest and vanilla.

Sift together the flour, baking powder and salt. Using a rubber spatula, stir in $^1/_3$ of the flour mixture. Then fold in $^1/_2$ of the milk. Repeat this procedure, alternating dry and liquid ingredients and ending with the flour.

Gently pour the mixture into the pan and smooth the top of the batter with a rubber spatula. Bake for 30–35 minutes, or until a cake tester comes out clean. While the cake is baking, prepare the coconut icing.

In a small saucepan, heat the cream, the remaining 3 Tbsp. (45 mL) butter, brown sugar and coconut over medium-high heat. Bring the mixture to a boil, stirring often.

Remove from the heat. When the cake is done, remove it from the oven and evenly spread the coconut icing over the top of the cake. Heat the broiler to high. Place the cake under the broiler until the icing is golden and bubbly, about 2–3 minutes. Place the cake on a rack and allow to cool completely before removing from the pan.

tip: **Crème Fraîche**

When making crème fraîche, choose a brand of whipping cream with a high percentage of milk fat and check the products are not near their expiry dates. Choose a glass bowl or jar and whisk together 2 cups (475 mL) whipping cream and $^1/_2$ cup (120 mL) buttermilk. Cover loosely and let stand unrefrigerated overnight in a warm spot. It should appear thick, like yogurt or sour cream, around the top. If the consistency is not correct, leave it for another 12 hours. Cover with plastic wrap or seal the jar tightly and refrigerate for up to 3 weeks. **–GM**

Menah's Layered Chocolate Cake
with Peppermint Icing

Serves 12

My mother's time in the kitchen when we were growing up was often spent trying to entice us to eat a balanced diet. This dark, moist chocolate cake handed down from my grandmother was enough to do just that. I've taken the liberty to "embellish the rose" by layering it with ganache.

2/3 cup	brown sugar, packed	160 mL
2/3 cup	sifted Dutch cocoa	160 mL
1 cup	boiling water	240 mL
1/2 cup	unsalted butter, room temperature	120 mL
2 cups	brown sugar, packed	475 mL
2	egg yolks	2
2 tsp.	baking soda	10 mL
1 cup	sour milk	240 mL
2 cups	flour, heaping	240 mL
2	egg whites	2
1 cup	whipping cream	240 mL
12 oz.	good-quality semisweet chocolate, cut in small pieces	340 g
1 recipe	Peppermint Icing	1 recipe

Preheat the oven to 375°F (190°C). Butter and flour an 11-inch (28-cm) springform cake pan.

Place the 2/3 cup (160 mL) brown sugar, cocoa and boiling water in a saucepan and bring to a boil over medium heat. Simmer for 5 minutes. Pour into a mixing bowl and stir constantly for 3–5 minutes to smooth out the lumps. Pour into another mixing bowl and cool for 15 minutes.

Beat together the butter, 2 cups (475 mL) brown sugar and egg yolks until smooth. Add to the cooled cocoa mixture. Dissolve the baking soda in the sour milk. Stir into the mixture.

Sift the flour into the mixture. Beat the egg whites until stiff. Fold into the mixture, mixing just until no lumps remain. Spoon into the prepared baking pan and bake for 55 minutes, or until the centre of the cake springs back when pressed. Cool in the pan on a rack.

Bring the cream to a boil, then remove from the heat. Add the chocolate. Let sit until the chocolate is melted, 3–4 minutes. Whisk until smooth. Cool.

Cut the cake in 3 equal layers. Spread the ganache between the layers. Ice the top and sides with peppermint icing in big swirls.

Peppermint Icing

Ices one 11-inch (28-cm) cake

2 cups	soft unsalted butter	475 mL
3 1/2 cups	sifted icing sugar	840 mL
2	egg yolks	2
1/4 tsp.	peppermint extract	1.2 mL

Beat the butter until light-coloured and fluffy, gradually adding the icing sugar. Beat in the egg yolks and peppermint extract.

tip: **Rolling Stiff Dough into Logs**

> When working with stiff doughs that have to be rolled into logs, roll the pieces lightly, stretching them slightly. Let them rest for a few minutes, then come back and roll them again. Continue the process until the dough is rolled to the right length. Using this slow process will prevent the dough from tearing or becoming uneven. **–KB**

Chocolate Truffle Cake

Serves 12

Chocolate times three! This is the cake I request for my birthday every year. It's moist and chocolatey and keeps very well. You can make it a day ahead or pre-make and freeze the cake itself or even freeze it with the truffle filling already added. I use simple syrup to moisten the layers as I assemble it. You could also use a favourite liqueur or even a thin layer of good-quality jam or preserve.

For the simple syrup:

1 cup	sugar	240 mL
1 cup	water	240 mL

Combine the sugar and water in a small saucepan and bring it to a boil over medium-high heat. When the sugar has dissolved (almost immediately), remove the syrup from the heat. Let it cool. This mixture will keep indefinitely if you keep it in a tightly covered glass jar in the refrigerator.

For the cake:

8	eggs	8
1 1/2 cups	sugar	360 mL
1 3/4 cups	flour	420 mL
3/4 cup	cocoa	180 mL
2 tsp.	baking powder	10 mL
1/2 tsp.	salt	2.5 mL

Preheat the oven to 325°F (165°C). Grease and flour a 9 1/2-inch (24-cm) springform pan.

In a large metal bowl, whisk together the eggs and sugar. Place the bowl over a pot of simmering water. The bottom of the bowl should not touch the water. Whisk the mixture until all of the sugar has dissolved, approximately 7–8 minutes.

Remove the bowl from the heat and beat the mixture with an electric mixer until it is light and foamy and 2 $1/2$-3 times its original volume.

In a separate bowl, sift together the flour, cocoa, baking powder and salt. Fold the dry ingredients gently into the egg and flour mixture until it is well incorporated. Pour the batter into the prepared pan and bake for 35–40 minutes. The cake should spring back when pressed gently. Remove to a rack to cool. When it has cooled, loosen the cake by running a knife between the edge of the cake and the pan. Remove it from the pan and set aside.

For the truffle filling:

3/4 lb.	butter	340 g
1/4 cup + 3 Tbsp.	sugar	105 mL
3	eggs	3
1 tsp.	vanilla extract	5 mL
1 3/4 cups	bittersweet chocolate, melted and still warm	420 mL

Using an electric mixer, cream the butter and sugar in a medium bowl. Add the eggs one at a time, beating for 1 minute after each addition. Add the vanilla and melted chocolate, and beat until the chocolate has been incorporated. Set aside.

(continued on following page)

DEB CONNORS

For the ganache:

1 cup	whipping cream	240 mL
2/3 cup	clarified butter (page 139)	160 mL
1 Tbsp.	corn syrup	15 mL
1 cup	bittersweet chocolate, finely chopped	240 mL

In a medium saucepan combine the cream, clarified butter and corn syrup. Bring the mixture to a boil over medium-high heat. Place the chocolate in a medium bowl. Remove the cream mixture from the heat and pour it over the chocolate, stirring until the chocolate has melted and the mixture is smooth. Place plastic wrap directly on top of the chocolate to prevent it from forming a skin. Leave this mixture to cool at room temperature until you are ready to use it.

To assemble the cake:

Carefully, using a sharp serrated knife and a gentle sawing motion, cut the cake horizontally in three layers. Place the bottom layer cut side up on a cardboard cake round or the removable bottom of the springform pan. Brush the layer with simple syrup and then spread with 1/2 of the truffle mixture. Repeat with the next layer, simple syrup and the remaining truffle mixture. Cover the cake with the top layer. Press down gently and evenly over the surface of the cake. Place the cake in the fridge until the truffle mixture has set, approximately 1 hour.

Remove the cake from the fridge. Place it on a wire baking rack over a cookie sheet to catch any ganache that runs down the sides of the cake. Pour the ganache on the cake and use a metal spatula to spread an even layer over the top and sides. Place the cake in the fridge for 25-30 minutes to set the ganache. Cut and serve.

Chocolate,
Cinnamon and
Pumpkin Seed
Cookie Brittle
(page 180)

Rhubarb Coconut
Cake (page 184)

Menah's Layered
Chocolate Cake with
Peppermint Icing
(page 188)

Rick's Granola Bars

I love to visit my friend Rick; he always has treats in his house. He bakes old-fashioned favourites like banana bread and his grandma's shortbread cookies. I really enjoy his granola bars, and it is a simple recipe to vary according to your taste. You can substitute other dried fruits and nuts, for instance, keeping the same proportions. I have thrown in pumpkin seeds to replace Rick's sunflower seeds.

Makes 24 bars

2/3 cup	unsalted butter,	160 mL
1/2 cup	brown sugar, packed	120 mL
1/2 cup	corn syrup	120 mL
2 tsp.	vanilla extract	10 mL
3 cups	rolled oats	720 mL
1/2 cup	unsweetened coconut flakes	120 mL
1/2 cup	sun-dried cranberries	120 mL
1/2 cup	chopped dried apricots	120 mL
1/2 cup	hulled pumpkin seeds, toasted	120 mL
1/2 cup	chopped pecans, toasted	120 mL
1/4 cup	sesame seeds, toasted	60 mL
1/4 cup	wheat germ	60 mL

Preheat the oven to 350°F (175°C).

In a large bowl, beat the butter, sugar, corn syrup and vanilla until smooth. Stir in the remaining ingredients.

Firmly pat the mixture into a well-greased 13 x 9-inch (33 x 23-cm) baking pan. Bake until golden, about 30–35 minutes. Place the pan on a rack and cool completely before cutting into 24 bars. Store the cut bars between layers of wax paper in a plastic container.

MARY MACKAY

Galette Nicola

Serves 6

Ginger snaps from England, lemon mousse from France and rhubarb from Canada—the flavour combination is divine. The inspiration for this dish came from Nicky Major; she was a pioneer of catering in Canada and I was lucky to be her chef for several years.

For the ginger snaps:

1/4 cup	unsalted butter	60 mL
1/4 cup	corn syrup	60 mL
1/3 cup	brown sugar	80 mL
7 Tbsp.	all-purpose flour	105 mL
1/2 tsp.	ground ginger	2.5 mL
	pinch salt	
1 tsp.	lemon juice	5 mL

Preheat the oven to 350°F (175°C).

Melt the butter, corn syrup and sugar together. Cool the mixture for 2-3 minutes. Stir in the flour, ginger, salt and lemon juice. Mix well until smooth. Cool the mixture until firm. Shape into 3/4-inch (2-cm) balls. Bake on a greased cookie sheet until golden brown and caramelized. Let cool 1 minute or so. Lift carefully off the pan with a thin spatula (metal works best). Cool on a rack.

For the rhubarb coulis:

3/4 lb.	rhubarb, thinly sliced	340 g
3/4 cup	granulated sugar	180 mL
1/2 cup	water	120 mL

Place the rhubarb, sugar and water in a small pot over medium-low heat and cook until the rhubarb is very soft, approximately 12-15 minutes. Cool. Purée in a food processor until smooth. Add a little water if necessary to make a thin sauce.

MARGARET CHISHOLM

For the lemon mousse:

3	eggs	3
2	egg yolks	2
1/2 cup	lemon juice	120 mL
1 Tbsp.	lemon zest	15 mL
2/3 cup	granulated sugar	160 mL
1/4 cup	butter	60 mL
2/3 cup	heavy cream, whipped	160 mL

Combine the eggs, egg yolks, lemon juice, lemon zest and sugar in a bowl. Place the bowl over a pot of simmering water. Whisk the mixture occasionally and cook until it's very thick. Whisk in the butter. Cool to room temperature. Fold in the whipped cream.

To assemble the galette:

	icing sugar	
1 pint	strawberries, sliced	500 mL

Place one ginger snap on a dessert plate. Top with a spoonful of lemon mousse. Top with another ginger snap. Repeat. Dust with icing sugar and spoon a few tablespoons (15 mL) of rhubarb coulis around the plate. Garnish with sliced strawberries.

quick bite: **Espresso Ricotta**

Dissolve 1/4 cup (60 mL) sugar in 2 Tbsp. (30 mL) of espresso. Add 1 tsp. (5 mL) of vanilla extract. Beat 1/2 cup (120 mL) whipping cream to soft peaks and fold into 2 cups (475 mL) whole-milk ricotta cheese. Swirl in the espresso and spoon into martini glasses. Dust with cocoa and serve.

–LS

Golden Fruitcake

Makes one
10-inch (25-cm)
cake

I actually like fruitcake, because my mother is such a fantastic baker. Her holiday repertoire includes heirloom recipes, her own classics and dark and light fruitcakes. The light cakes, studded with almonds and pale raisins, became my base for the next generation of fruitcake and a new tradition. This cake is bright with citrus and apricot, subtly nutty and moist. Serve with a steaming hot cup of tea, a glass of dessert wine or bubbly.

1/2 lb.	raisins (golden, Thompson, Muscat or a mixture)	227 g
1/2 lb.	dried apricots	227 g
1/2 lb.	candied peel (lemon, orange, citron or a mixture)	227 g
1 cup	apricot or orange brandy or almond liqueur	240 mL
3/4 cup	unsalted butter, at room temperature	180 mL
1 cup	sugar	240 mL
1/2 cup	whipping cream	120 mL
4	eggs	4
2 1/4 cups	flour	535 mL
1/2 tsp.	baking powder	2.5 mL
1 Tbsp.	almond extract or bitter almond extract (available at specialty stores)	15 mL
1 tsp.	vanilla extract	5 mL
1 cup	whole unblanched almonds, coarsely chopped	240 mL
1/2 cup	light corn syrup	120 mL

Chop or scissor-cut the large raisins and apricots. Spraying the knife or scissor blades with vegetable spray will prevent sticking.

GLENYS MORGAN

Moist, sticky fruit is a must for fruitcakes. Soak the raisins, candied peel and apricots overnight (or for several days) in the brandy or liqueur of choice. For a quick method, poach the fruit with the brandy over heat, until the fruit is sticky and the excess liquid nearly gone. Cool. The quick-glazed fruit will be more sticky than juicy.

In a large mixing bowl, cream the butter and sugar until pale. Add the cream and eggs, beating to incorporate (it may appear curdled like buttermilk). Blend 2 cups (475 mL) of the flour with the baking powder; fold or slowly mix it into the creamed mixture. (To double or triple the recipe, use the electric mixer, transferring the batter to a large bowl to fold in the fruit and nuts.) Add the almond and vanilla extracts.

Toss the fruit and nuts with the remaining 1/4 cup (60 mL) of flour. Use a large wooden spoon or rubber spatula to fold in the floured fruit and nuts. Generously butter or spray the inside of a 10-inch (25-cm) tube pan (kugelhopf). Spoon the batter into the pan, filling it 2/3 full, and levelling with a rubber spatula. Tap the pan on the counter to eliminate air bubbles in the batter. If the pan is very full, extend the height of the pan with a collar of foil wrapped around the pan; tie it in place with butcher's twine.

Preheat the oven to 275°F (135°C). Place the cake in the centre of the oven and bake for approximately 90 minutes. (Note: You can divide the batter between two 6-inch/15-cm tube pans and bake for approximately 60 minutes, or bake in an 8-inch/20-cm square fruitcake pan for approximately 2 hours.) Use a bamboo skewer to test the thickest part of the cake. If the cake is done, the skewer will be dry and clean; if the skewer is sticky, the cake is not done. When done, remove from the oven and cool on a rack. Cool until warm, then brush the surface with corn syrup. Unmould the cake when it's completely cool.

Wrap the cooled cake in plastic wrap with an exterior wrap of foil to retain moisture. It will be ready to eat after 1 day. The cake may be refrigerated for a few weeks or frozen for up to several months. The cake can periodically be unwrapped and brushed with your favourite liqueur or brandy as it ages.

Chèvre Panna Cotta with Poached Fruits

Serves 6

I was asked to design a fundraising dinner for the Dance Foundation Rooftop Deck Project, to be held in the magnificent secluded garden of pre-eminent architect Arthur Erickson. The setting inspired me to create a dinner of contrasts, to complement a garden focused on foliage, colour and texture. This is the dessert we served.

1	envelope gelatin	1
1/4 cup	water	60 mL
1 1/2 cups	whipping cream	360 mL
1 1/2 Tbsp.	honey	22.5 mL
1/2 tsp.	vanilla extract	2.5 mL
5 oz.	soft chèvre	140 g
2 cups	apple juice	475 mL
1 cup	mixed dried fruit (use any combination of dried fruit—apricots, prunes, cherries, blueberries, currants, raisins or figs will all contribute something special)	240 mL
1	cinnamon stick	1
3	whole cloves	3
6	pods cardamom, bruised	6

Sprinkle the gelatin over the water and allow to soften for 10 minutes. Place the cream and honey in a small pot and heat to a simmer. Remove from the heat. Add the gelatin and stir until dissolved. Add the vanilla. Place the chèvre and warm cream in a food processor and process until smooth. Pour the mixture into 6 5-oz. (140-mL) ramekins. Chill for 2 hours or more.

Place the apple juice in a non-reactive (stainless steel, glass or enamel-lined) saucepan. Add the dried fruit and spices. Place over medium heat and bring to a very gentle simmer. Turn the heat down to low, cover and allow the fruit to poach for 30 minutes.

Strain the fruit, reserving the juice. Remove the whole spices and discard. Return the juice to the saucepan and simmer gently until reduced to a light syrup consistency.

To serve, run the tip of a knife around the edge of the ramekins to loosen the custard. Unmould onto a dessert plate. Arrange some poached fruit around the outside of each custard. Drizzle the fruit with some of the syrup and serve with a piece of Maple Hazelnut Shortbread (page 181).

---

quick bite: **Gus and Mary's Strawberries, Balsamic Vinegar and Cream**

Two friends of mine presented this dessert after a dinner on a hot summer's night in Toronto. It was simple, elegant, perfect and it has stayed with me ever since.

For each person, they had two small, shallow and narrow glass dessert dishes. One was filled with whipping cream; the other was filled with a good-quality balsamic vinegar. In the middle of the table they put a plate of intact, ripe, summer strawberries. We each picked up a strawberry by the hull and dipped it into the balsamic vinegar, then into the cream, then into our mouths. All I could say was "Wow!"

The acid reaction of the vinegar made the cream cling to the berries and because the dishes were narrow, we had enough cream and vinegar to dip to our hearts' content. When the berries started coming to an end, I mashed a few berries with the cream, mixed in the balsamic vinegar and sipped the heavenly concoction. **−KB**

Rosemary Mascarpone Cheesecake with Peppered Raspberry Coulis

Makes one
9-inch (23-cm)
cheesecake

From potpourris to aromatherapy, a new look at the traditional uses of herbs has inspired cooks to do the same. Rosemary, thyme, lemon verbena, basil–for some time used only for savoury dishes–are making their way into sweets, as they did hundreds of years ago. Some, like the cooks of Provence, have known it's the thing to do all along.

2	6-inch (15-cm) sprigs fresh rosemary	2
1 cup	whipping cream	240 mL
1 lb.	cream cheese, at room temperature	455 g
1 lb.	mascarpone cheese	455 g
1 cup	sugar	240 mL
4	extra-large eggs	4
1	lemon	1
1 recipe	Peppered Raspberry Coulis	1 recipe

Infuse the flavour of the rosemary into the whipping cream by heating gently in a small saucepan. Scald and remove from the heat to cool. The cream will absorb the rosemary flavour while you prepare the rest of the ingredients. (Steeping the rosemary sprigs in whipping cream overnight in the refrigerator is another method.) When cool, strain and discard the rosemary; reserve the whipping cream until needed.

Preheat the oven to 325°F (165°C). Prepare a 9-inch (23-cm) round cake pan by lining the bottom with a parchment or wax paper circle cut to fit.

Use the mixer or food processor to combine the cheeses, blending until smooth. Add the sugar and beat until it begins to dissolve. Add the eggs one at a time, beating well after each addition.

Use a zester to remove the lemon zest. Juice the lemon and add both zest and juice to the mixture. Add the infused whipping cream and beat again.

GLENYS MORGAN

Pour into the prepared pan. Level the batter by spinning the pan on the counter. This will even out the cake without compacting it. Place the pan in a larger baking pan. Fill the outer pan with enough boiling water to come 2/3 of the way up the side of the cake pan. Place in the preheated oven and bake for 45 minutes, or until firm but lightly set.

Cool the cake, then chill it to finish setting the creamy texture. This cake has better flavour if made a day ahead, allowing the flavours to mature, but it can be served as soon as it's well chilled. Run a knife around the cake to release the sides. Place a dinner plate over the cake and turn the cake onto the plate. Peel off the parchment or wax paper. Place 2-inch (5-cm) slices bottom side down on individual serving plates. Wipe the knife between cuts to keep the edges neat. Serve with a spoonful of sauce drizzled on the plate.

Peppered Raspberry Coulis

Makes approximately 1 cup (240 mL)

2	packages frozen raspberries in syrup	2
1 Tbsp.	pink peppercorns in brine, drained	15 mL
2 Tbsp.	rum	30 mL

Thaw the raspberries. Press the raspberries through a sieve or food mill to remove the seeds. Scrape the underside of the sieve or food mill to collect the thicker purée. Combine the purée with the peppercorns and rum. Chill until needed.

If the raspberry mixture is thinner than desired, reduce gently over low heat in a small saucepan. Add the peppercorns and rum to the reduced coulis.

Dark Chocolate Pâte

Serves 8 to 10

This is a very simple dessert that you can make a day ahead. It also freezes very well. For a simple mould you can use small paper cups set inside muffin tins (this will make 8-10 individual desserts). You can also use a single mould lined with parchment paper and slice the dessert with a hot knife.

1 cup	whipping cream	240 mL
16 oz.	dark chocolate, chopped into small pieces	455 g
1/2 cup	clarified butter (page 139)	120 mL
1/3 cup	icing sugar	80 mL
4	egg yolks	4
	whipped cream	
	fresh strawberries	

Scald the cream in a small saucepan over medium-high heat. At the same time melt the chocolate in a medium bowl over a pot of simmering water. Do not let the bottom of the bowl touch the water.

Remove the melted chocolate from the heat and alternately whisk 1/2 of the clarified butter and 1/2 of the scalded cream into the chocolate. Add the remaining butter and then the remaining cream.

In a small bowl whisk the icing sugar into the egg yolks until just blended. Add this mixture to the melted chocolate and stir until it is just combined. Pour the chocolate into the mould or moulds and refrigerate until set. Use any mould you have, such as a standard loaf pan. Cut the pâte into whatever shape you like—squares, rectangles or triangles.

To serve, turn the pâte out of the mould onto a plate. Top each serving with whipped cream and fresh berries.

DEB CONNORS

Toasted Almond Torte
with Burnt Orange Caramel Sauce

I grew up with Mom's walnut squares with shortbread undersides and her butter tarts. (Here's some trivia I learned recently: those perfectly gooey tarts are Canadian.) This torte has the shortbread base of walnut squares; the top layer is a cousin to the butter tart. I made this orange and toasty almond combination to pair with champagne or late harvest wines, so it's not as sweet as some.

Makes one
10-inch (25-cm)
torte

For the base:

1 1/4 cups	flour	300 mL
1/4 cup	sugar	60 mL
6 Tbsp.	unsalted butter, frozen	90 mL
1	egg	1
1 tsp.	almond extract or bitter almond extract (available at specialty stores)	5 mL
1 Tbsp.	grated orange zest	15 mL

Preheat the oven to 350°F (175°C). Spray a 10-inch (25-cm) removable bottom tart pan with vegetable spray.

Combine the flour and sugar in the food processor. Cut the frozen butter into chunks and distribute evenly over the flour in the food processor. Whisk together the egg, extract and zest. Add to the food processor and pulse to begin working the crust together. Do not overmix; the dough should look like large shortbread crumbs.

Turn the crust mixture out into the tart pan. (Set aside the food processor bowl without washing it to make the topping.) Use your fingers to loosely spread the crumbly mixture evenly over the pan base, patting into a loose even layer. To give the outside edge of the torte a professional-looking finish, use your fingertips to tamp down the crust in the pan's fluted edge. The finished base should be completely flat and loosely patted down. Packing the base will make it too hard. Chill the base until the topping is prepared.

(continued on following page)

GLENYS MORGAN

For the topping:

2/3 cup	whole unblanched almonds	160 mL
1/2 cup	sugar	120 mL
1 cup	unsalted butter, chilled	240 mL
1	egg	1
1	egg yolk	1
1 Tbsp.	almond extract	15 mL
1 tsp.	vanilla extract	5 mL
1 cup	sliced toasted almonds	240 mL

Place the whole almonds in the food processor with the sugar. Grind to a coarse powder. Cut the butter into chunks and add to the food processor. Blend until creamy and add the egg, egg yolk and extracts. Process until just blended.

Use a spatula to remove the almond topping from the food processor bowl, mounding it in the centre of the chilled crust. Spread the topping with the spatula, each time in a single stroke from centre to pan edge, wiping the spatula on the rim. Smooth, even strokes prevent the sticky topping from lifting the crust. Repeat from the centre to the edge until the torte base is covered.

Sprinkle the almond topping with an even layer of sliced toasted almonds and gently pat onto the surface. Once the torte is baking, it will melt into a smooth even layer with the almonds settling on top. Place the torte pan on a baking sheet to avoid burning oil in the oven; as it bakes, the melting top may drip through the removable bottom.

Bake the torte until it's browned and the topping is set and firm, about 45 minutes. Cool on a baking rack. Remove the outside rim of the pan and cut into 16 2-inch (5-cm) wedges. Serve each wedge with a spoonful of Burnt Orange Caramel Sauce drizzled on the plate.

Burnt Orange Caramel Sauce

Makes 1 cup (240 mL)

2	oranges	2
1 cup	sugar	240 mL
	few drops lemon juice	
1/2 cup	water	60 mL
1 Tbsp.	unsalted butter	15 mL
1/4 cup	whipping cream	60 mL

Have everything you need for the sauce at hand; it comes together so quickly it's best to give it your full attention. Remove the zest from the oranges in long thin strips with a zester (do not use a citrus rasp) or peeler. Juice the oranges and set aside. Combine the zest, sugar, lemon juice and water in a small saucepan. (The lemon juice helps avoid crystallization.) Heat to boiling over medium-high heat.

As the water and sugar melt, they will foam in large frothy bubbles. Never stir the caramel, but rotate the pan so the liquid moves the sugar around for even melting. Watch the liquid turn from clear to gold; from this point it will quickly turn to amber and burn. Remove the pan from the heat as soon as it appears dark gold. Place a mesh strainer over the pot (this prevents spattering) and add the orange juice.

Add the butter to the pot and return to the heat. Add the whipping cream and stir until any remaining lumps dissolve. Strain to remove the zest. Store in the refrigerator and warm like a baby bottle—in the microwave or in a simmering pan of water.

tip: **Sugared Mint Leaves**

Dip small sprigs of mint leaves with stem attached into beaten egg white and then into a little extra-fine sugar. Leave them to day for an hour or so on parchment paper. **–DC**

GLENYS MORGAN

Inspiration is what you make of it.

MY INSPIRATIONS COME FROM many people, places and experiences, starting at the skirts of my Russian gramma. Vivid memories of watching her cook from her large farmhouse kitchen still bring back the smell of her deep spice drawer, from which she pinched and hand-scooped, but never measured, the essences of her cooking.

Life deals out many cards that can either be played or traded. One of my cards was being the prep cook in the house. By virtue of being an only child, my card had to be played and I was delegated cook when my mother worked.

My interest flourished when I began a career as a flight attendant with CP Air. I was fortunate enough to be awarded international trips, and the markets of the world became my inspiration. Amsterdam opened my eyes to how *frites* should really be served—in a paper cone adorned with a generous dollop of frite sauce (mayonnaise)—and the smelly pickled herring so loved by the Dutch, always accompanied by a chilled Amstel. I also experienced *uitsmidter*, stacked and fried in many different ways, which were the inspiration in later years for fantastic egg concoctions.

From the noodle houses in Hong Kong came more ideas for interesting foods and ways of preparing them. The cracking of the clay from the fabulous clay-baked chicken revealed a tender, succulent, slow-cooked treat that was protected from drying out by the cooking method. Keeping an open mind, however, wasn't always easy. My most uninspired experience came upon my graduation in Tokyo, Japan. I was showered with gifts and honoured with a formal Japanese dinner. With a gleeful expression, our server presented us with a pulsating sea urchin surrounded by crisply grilled snake (this pretty much dampened my eagerness for culinary adventure for some time to come).

CAREN MCSHERRY-VALAGAO

CAREN MCSHERRY-VALAGAO

Travel contributes enormously to the creativity of food educators and chefs. Our global table integrates foods and cultures around the world. Because I spent a great deal of my twenties and thirties travelling and attending culinary schools and professional kitchens worldwide, my inspirational window was opened wide. I tasted a wide range of foods, from a simple but gutsy and flavourful Italian street sandwich called *muffaletta* to the ridiculously decadent zucchini blossom stuffed with an entire truffle in the south of France at Moulin de Mougin.

These experiences became the foundation of my cooking career and philosophy of cooking—creating dishes that cater to all tastes, from peasant to prince.

Inspiration is what you make of it. On these pages we have given you a broad palette of flavours and ingredients; your challenge is to interpret these ideas and make creations of your own. But— most important—remember that inspiration comes from loving what you do. Without the love and passion, you cannot be inspired.

CAREN MCSHERRY-VALAGAO

Gramma Koch's Apple Strudel

Serves 8 to 10

My grandmother was a wonderful cook. She used to make incredible strudel dough that would stretch the entire length of the farmhouse table. I was mesmerized as I watched her coax and roll a little ball of dough into a paper-thin blanket that could envelope a human, to say nothing of the apples it was meant for. If you can talk yourself into trying the dough, the rest is easy. Settle for a small, side-table-size dough and leave the farmhouse-table size to the German grammas that created it. Serve the strudel warm or at room temperature with ice cream or whipped cream.

1	large egg	1
1 Tbsp.	corn or grapeseed oil	15 mL
1 Tbsp.	water	15 mL
1 cup	unbleached flour	120 mL
1/3 cup	unsalted butter, melted	80 mL
5	apples (Granny Smith or Golden Delicious), peeled, cored and thinly sliced	5
1/2 cup	sugar	120 mL
1/2 cup	toasted slivered almonds	120 mL
2/3 cup	raisins or currants	160 mL
1 tsp.	ground cinnamon	5 mL
4	soda crackers, crushed	4

Lightly beat the egg. Add the oil and water. Place the flour in a mixing bowl. Pour the egg mixture over the flour. Mix until blended. Turn the dough out onto a lightly floured work surface and knead until smooth and no longer sticky. Place the dough on a marble slab or smooth work surface. Invert the bowl over the dough and let the dough rest for at least 5–6 hours. (This is the secret of getting it to stretch; you are relaxing the gluten.)

When the dough has rested, preheat the oven to 350°F (175°C). Drape a cotton cloth that is at least 3 x 3 feet (1 x 1 m) over your rolling surface. Sprinkle lightly with flour. Place the rested dough in the centre of the sheet and begin rolling. It will take time and coaxing, but the dough will roll out to at least 24 x 24 inches (60 x 60 cm). Not quite a farmhouse table but good enough! The dough must be very thin.

Leaving a $1/2$-inch (1.2-cm) border around the edge, brush the dough with the melted butter. Spread the sliced apples evenly over top. Sprinkle the sugar evenly over the apples, then add the almonds, raisins or currants, cinnamon and soda crackers. Roll the strudel up, jelly-roll style. Roll it directly onto a large baking sheet. Brush the top with melted butter and bake for about 45 minutes.

tip: **Frozen Citrus Zest**

My kitchen is never without lemons and limes, but just in case I need citrus zest, when I've juiced an orange or lemon or lime with perfect skin, it goes in a bag in the freezer. The essential oils are drawn to the surface in freezing and when they're frozen solid, they're easy to zest and always handy. **–GM**

CAREN MCSHERRY-VALAGAO

Best Breakfast (or Dessert) Waffles

Serves 4

I am a breakfast fanatic, although I rarely have time to linger over this meal. When I do, the waffle iron comes out, the newspapers are piled high and a second pot of coffee is *de rigueur*. My husband is the expert in the kitchen when it comes to breakfast. With a little coaching from me, he has come up with the best waffles I've had yet. These also make a great dessert with fresh berries and some top-quality ice cream!

2	eggs, separated	2
1 3/4 cups	milk	420 mL
1/4 cup	melted butter, slightly browned	60 mL
1/2 tsp.	vanilla extract	2.5 mL
1 3/4 cups	flour	420 mL
4 tsp.	baking powder	20 mL
2 1/2 Tbsp.	sugar	37.5 mL
1 tsp.	sea salt	5 mL
1/2 cup	grated apple	120 mL

Preheat the waffle iron. Beat the egg yolks, milk, butter, and vanilla together. Sift the flour, baking powder, sugar and salt into a bowl. Mix the liquid into the dry ingredients, stirring just until the large lumps disappear. Stir in the apple.

Beat the egg whites until stiff but not dry. Fold them into the batter. Pour into the hot waffle iron, following the manufacturer's instructions. Serve hot, with lots of butter and warm real maple syrup.

Profiteroles with Orange Mascarpone Cream and Chocolate Crème Anglaise

I've always loved the combination of oranges and dark chocolate. This is a lighter presentation, with the orange flavour being incorporated into the mascarpone and the anglaise carrying the chocolate. The puffs must be piped and baked as soon as the batter is completed, but you can make all of the components a day ahead. Store the profiteroles in an airtight container; if they get soggy you can re-crisp them in a preheated 300°F (150°C) oven. The profiteroles also freeze well. They can be filled and refrigerated several hours prior to serving.

Serves 6 to 12, depending on the serving size

1/2 cup	water	120 mL
3 Tbsp.	butter	45 mL
2 tsp.	sugar	10 mL
1 Tbsp.	minced orange zest	15 mL
	pinch salt	
1/2 cup + 2 Tbsp.	all-purpose flour	150 mL
2-3	large eggs	2-3
1 recipe	Orange Mascarpone Cream (page 212)	1 recipe
1 recipe	Chocolate Crème Anglaise (page 213)	1 recipe
	icing sugar	
	lemon balm or mint leaves	

Preheat the oven to 400°F (200°C).

In a medium saucepan combine the water, butter, sugar, orange zest and salt. Bring to a boil over high heat. When the mixture is boiling rapidly, add all the flour at once and stir with a wooden spoon. Turn the heat to medium and continue to stir for 2 minutes, or until the mixture pulls away from the sides of the saucepan. Remove from the heat and turn the dough into the bowl of an electric mixer. Let cool for a few minutes.

(continued on following page)

DEB CONNORS

Using the paddle attachment, beat the mixture at medium speed for 2 minutes. Add 1 egg and beat for 2 minutes. Repeat with the second egg. Lift the paddles; the dough should have a smooth and silky appearance, and it should form into a soft peak. If the batter is too stiff, add the last egg and beat for 2 more minutes.

Using a piping bag with a plain 1/2-inch (1.2-cm) tip, pipe the batter in 2-inch (5-cm) circles onto a lightly buttered baking sheet, leaving 3 inches (7.5 cm) of space between each profiterole. You will get 10-12 profiteroles. Bake for 5 minutes. Reduce the heat to 350°F (175°C) and bake for another 15-20 minutes, until the profiteroles are firm to the touch and a light golden brown. Remove them to a rack to cool, piercing the side of each one with a sharp knife to allow the steam to escape.

To assemble the profiteroles, slice the top off each profiterole and set aside. Spoon in the Orange Mascarpone Cream. Place a small pool of Chocolate Crème Anglaise in the centre of a small plate. Place the profiterole in the centre of the plate, drizzle with anglaise and replace the top. Dust with icing sugar and garnish with a lemon balm or mint leaf.

Orange Mascarpone Cream

Makes 3 cups (720 mL)

1 cup	freshly squeezed orange juice	240 mL
3 Tbsp.	sugar	45 mL
3/4 cup	mascarpone cheese	180 mL
1 cup	whipping cream	240 mL
1 Tbsp.	Grand Marnier (optional)	15 mL

Combine the orange juice and sugar in a small sauté pan over high heat. Cook until it's reduced to 2 Tbsp. (30 mL) of syrup, watching carefully so it does not burn. Remove from the heat and cool.

In the bowl of an electric mixer combine the cooled orange syrup and the mascarpone. Using the whip attachment, beat at medium speed for 2 minutes until well mixed. Add the whipping cream and Grand Marnier. Turn the speed to high and beat until the mixture is stiff. Cover and refrigerate.

Chocolate Crème Anglaise

Makes ³/₄ cup (180 mL)

2	egg yolks	2
2 Tbsp. + 2 tsp.	sugar	40 mL
¹/₃ cup	milk	80 mL
¹/₃ cup	whipping cream	80 mL
¹/₃	vanilla bean, halved lengthwise	¹/₃
1 oz.	bittersweet chocolate, melted	28 g

In a heatproof medium bowl combine the egg yolks and sugar. Whisk until thick, 3-5 minutes.

In a small heavy saucepan, combine the milk, whipping cream and vanilla bean. Scald over medium-high heat. Very slowly pour the scalded cream into the egg mixture, whisking constantly. Remove the vanilla bean.

Put the bowl over a pot of simmering water and stir the mixture with a wooden spoon until it starts to thicken, about 5 minutes. It should coat the back of the spoon. Remove from the heat and stir in the melted chocolate. Cool the anglaise and refrigerate.

---

tip: **"Stacking" Food the Easy Way**

Kitchen stores and gourmet retailers sell plain biscuit cutters and metal rings in many sizes. My favourite size is 2-3 inches (5-7.5 cm) wide and at least 1 inch (2.5 cm) deep. PVC pipe cut to size is a favourite for shaping "cakes" in the professional kitchen, but metal goes to the pan or oven and can be used for chilled presentations as well. Use them for crab cakes, moulding individual presentations of Death by Chocolate or your own mini "stacked" inspiration! –GM

Chocolate Peanut Butter Ice Cream

**Makes about
4 cups
(950 mL)**

Last summer I purchased an electric ice-cream maker and have been having a lot of fun with it. My husband, John, also enjoys using it to prepare ice cream, sorbets and frozen drinks. To keep things simple, look for a machine that does not require hand cranking or salt for freezing. My favourite chocolate bar, the peanut butter cup, inspired this recipe. I recommend serving your guests a scoop of the ice cream on top of a peanut butter cup.

2 cups	whole milk	475 mL
5	large egg yolks	5
6 Tbsp.	sugar	90 mL
1 tsp.	vanilla extract	5 mL
1 cup	whipping cream	240 mL
	pinch fine sea salt	
1/2 lb.	milk chocolate, melted	227 g
3/4 cup	crunchy-style peanut butter	180 mL

Place the milk in a saucepan over medium heat and bring to a simmer. Place the egg yolks and sugar in the bowl of an electric mixer and beat on medium-high speed until pale and thickened, about 4 minutes. Beat 1/2 cup (120 mL) of the hot milk into the egg mixture. Stir the egg mixture into the remaining hot milk, and cook over low heat, stirring constantly, until it is thick enough to coat the back of a wooden spoon. Remove the saucepan from the heat and stir in the vanilla, cream and salt. Whisk in the melted chocolate. Strain the mixture into a clean bowl and cover the surface with plastic wrap. Refrigerate the custard for at least 2–3 hours, until very cold.

Transfer the chilled custard to the bowl of the ice-cream maker and follow manufacturer's instructions for freezing. In the last 5 minutes of freezing add the peanut butter to the bowl of the ice-cream maker, while the machine is churning. Transfer the ice cream to an airtight container and store in the freezer for about half an hour before devouring.

MARY MACKAY

THE INSPIRED PANTRY

With a well-stocked pantry, inspired dishes are only moments away. This pantry list contains ingredients that many recipes in this book have in common.

In the Baking Cupboard

Almond extract
Baking powder
Chocolate, bittersweet
Chocolate, milk
Cocoa powder, Dutch-process
Cornstarch
Corn syrup
Gelatin
Flour, all-purpose
Honey

Maple syrup
Molasses
Peppermint extract
Sugar, dark brown
Sugar, demerara
Sugar, granulated
Sugar, icing
Sugar, light brown
Sugar, maple
Vanilla, pure extract and whole beans
Yeast, dry active

Canned Goods

Anchovy filets
Artichoke hearts
Bamboo shoots
Capers in brine
Chilies, chipotle en adobo
Coconut milk
Curry paste, Thai green
Olives, black pitted
Olives, Kalamata
Sun-dried oil-packed tomatoes
Tapenade
Tomatoes, diced
Tomato paste

In the Freezer

Stock, Chicken
Peas

Sauces and Condiments

Chili sauce, Chinese
Chili sauce, Thai sweet
Fish sauce, Thai
Hoisin sauce
Horseradish
Kecap Manis
Mango chutney
Mustard, Dijon and whole grain
Oyster sauce
Sambal oelek
Soy sauce
Wasabi paste or powder
Worcestershire sauce

Dried Goods

Cornmeal, ground yellow
Noodles, Papardelle
Oatmeal
Pasta, penne
Rice, Arborio
Rolled oats, quick cooking
Wheat germ

Dried Fruits

Apricots
Cherries
Cranberries, sun-dried
Dates, pitted
Figs, black Mission
Mixed dried fruit
Raisins

Nuts and Seeds

Almonds, whole unblanched and slivered
Cashews
Coconut, unsweetened flaked
Hazelnuts
Kalonji seeds
Macadamia nuts
Peanut butter, crunchy
Pecan halves
Pine nuts
Pistachios, roasted unsalted
Pumpkin seeds, hulled
Sesame seeds, white and black
Walnut pieces

Herbs and Spices

Allspice, ground
Bay leaves
Cardamom, ground and pods
Chai, herbal spice
Chili powder
Chili powder, Pasilla
Cinnamon, ground and stick
Cloves, whole
Coriander, ground
Cumin, ground
Curry powder
Fleur de Sel
Ginger, ground
Garam masala
Nutmeg, whole
Oregano, Mexican
Paprika, sweet and smoked
Peppercorns, pink in brine
Peppercorns, Telicherry
Peppercorns, whole black
Porcini powder
Saffron threads
Sea salt, fine and coarse
Turmeric

Oils

Grapeseed
Olive, extra virgin
Sesame

Vinegars

Apple cider
Balsamic
Fig balsamic
Red wine
Rice
Sherry
White wine

Alcohol

Rum
Sake
Vermouth, white
Wine, red
Wine, white

Miscellaneous Stuff

Bread crumbs
Marshmallows
Parchment paper

MEAL INSPIRATIONS

We would like to think that this cookbook is one you'll come back to time and time again for inspiration in your cooking. To get you started, we've put together a few menu ideas.

Fiery Feast

Mussels with Peppers, Bamboo Shoots and Green Curry, page 15

Fiery Thai Beef with Herb Salad and Cooling Cucumber Pickle, page 134

Cumin Roasted Yams, *The Girls Who Dish*, page 127

Crispy Chocolate and Black Pepper Shortbread, *The Girls Who Dish: Seconds Anyone?*, page 175

Fiesta Mexicana!

Tomato, Cumin and Black Pepper Tortillas with Chèvre, page 26

Spicy Jumbo Prawns with Nutty Herb Sauce, *The Girls Who Dish: Seconds Anyone?*, page 26

Fresh Tomato and Chorizo Soup with Chipotle Cilantro Cream, page 86

Chocolate, Cinnamon and Pumpkin Seed Cookie Brittle, page 180

Eastern Inspirations

The Alfresco Luncheon: Simple and Elegant

A "Do" a Deux

Middle Eastern Semi-Mezze: Great for a Barbecue!

A Smart Little Wine Tasting

Thoughts for Buffets

And the beauty of it is that everything can be served at room temperature!

...and remember to Label Your Party Platters, page 69!

Sensational Seafood

Comfort Food Is Always in Fashion

The Whole Is Greater...
Than the Sum of Its Parts

INDEX
OF RECIPES BY CHEF

INDEX OF RECIPES BY CHEF

DEB CONNORS

MARY MACKAY

INDEX OF RECIPES BY CHEF

INDEX OF RECIPES BY CHEF

LESLEY STOWE

INDEX OF RECIPES BY CHEF

INDEX

INDEX

INDEX

INDEX

INDEX

INDEX

INDEX

INDEX

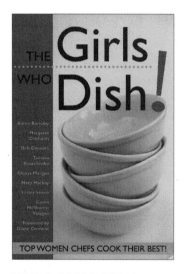
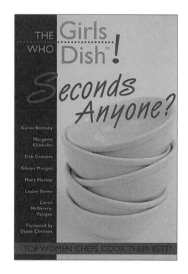